Essential Smart Growth Fixes for Rural Planning, Zoning, and Development Codes

ACKNOWLEDGMENTS

Principal author:

Kevin Nelson, AICP, Office of Sustainable Communities

Contributing content experts:

Chris Duerksen, Esq., Clarion Associates

Faith Ingulsrud, State of Vermont

Lee Nellis, FAICP, Round River Planning

Dru Schmidt-Perkins, 1,000 Friends of Maryland

Leslie Oberholtzer, Farr Associates

Reviewers:

Tim Davis, Sonoran Institute

Dave Mills, City of Ranson, West Virginia

Hugh Morris, National Association of Realtors

David Sears, U.S. Department of Agriculture Rural Development

Contributors and reviewers from U.S. Environmental Protection Agency:

Bicky Corman, Office of Policy

Andy Crossland, Office of Water

Matthew Dalbey, Office of Sustainable Communities

John W. Frece, Office of Sustainable Communities

Bonnie Gitlin, Office of Water

Melissa Kramer, Office of Sustainable Communities

Kellie Kubena, Office of Water

Lynn Richards, Office of Sustainable Communities

Megan Susman, Office of Sustainable Communities

Maureen Tooke, Office of Water

February 2012

CONTENTS

INTRODUCTION

Most rural communities want to maintain their rural character while also strengthening their economies. Many fast-growing rural areas are now at the edge of major metropolitan regions and face metropolitan-style development pressures. They seek to manage new growth in a way that promotes prosperity yet is sustainable over the long run. But even slow-growing or shrinking rural areas, which often suffer from faltering economies and population decline, might find that their growth management policies are not resulting in the prosperity they seek.

Fortunately, a variety of proven tools and strategies can help rural communities thoughtfully consider how and where to grow. For example, communities that want to maintain their rural character and economic vitality could decide to adopt mixed-use zoning for their Main Street buildings and commercial areas, policies to better manage stormwater runoff, and design requirements for complete, connected streets. Strategies like these are used in communities of all sizes around the country. Small towns and rural areas generally have fewer financial, technical, and staff resources to draw on in responding to development proposals and growth pressures than their urban and suburban counterparts. As a result, rural communities need to identify strategies that they are able to implement with their resources.

This publication provides a range of strategies organized around 10 chapters that focus on key issues that rural communities face. It is intended to provide smart growth policy options that communities can implement. These policies can help small towns and rural areas ensure that their development is fiscally sound, environmentally responsible, and socially equitable. This publication is a companion to *Essential Smart Growth Fixes for Urban and Suburban Zoning Codes.*[1] While many of the essential fixes from that document can be adopted in communities of any size, this publication provides additional

Seneca Falls, New York, has a thriving downtown with streets that are pleasant to walk along. Its "heritage area" designation preserves its history and attracts visitors.

options specifically for rural communities. This publication does not provide model codes; rather, it offers a range of options communities can consider implementing to make their development patterns more fiscally and environmentally sustainable.

Some rural parts of the United States do not engage in planning, zoning, or creating building codes. Since land use authority largely rests at the local level, local decision-makers have this prerogative. This document contains resources that can help rural communities along the spectrum of local land use controls.

With planning and zoning that supports their vision, rural communities can flourish and improve the quality of life for their residents, attract and support businesses, and provide new opportunities while protecting the way of life they cherish. This document identifies methods for getting the type of development that works best in a rural context.

---

1   EPA. *Essential Smart Growth Fixes for Urban and Suburban Zoning Codes.* 2009. EPA 231-K-09-003. http://www.epa.gov/smartgrowth/essential_fixes.htm.

SMART GROWTH IN RURAL AREAS

Smart growth development approaches benefit the economy, the environment, public health, and the community as a whole. In rural communities, smart growth strategies address the relationship between the land and the small towns and villages that support rural economies. Working agricultural lands, prairies, forests, and natural resource extraction historically drove the economy in many rural towns. Hamlets and villages grew as places to trade goods and services and as transportation hubs that connected the land-based economy to markets. Historically, these places were economic, civic, cultural, and social hubs. The villages had many of the characteristics that even today are important attributes of attractive, healthy places. Homes were within walking distance of stores and workplaces; land was used efficiently by clustering village-related uses in the village and keeping farms and other working lands as large swathes of land with little or no development to interfere with the economic uses.

The International City/County Management Association's *Putting Smart Growth to Work in Rural Communities* discusses trends affecting rural America today and how rural communities can use smart growth strategies to prosper. That publication suggests that if communities want to maintain their rural character, they should pursue three goals using smart growth approaches:

- Support the rural landscape by creating an economic climate that enhances the viability of working lands and conserves natural lands.

- Help existing places thrive by taking care of assets and investments such as downtowns, Main Streets, existing infrastructure, and places that the community values.

- Create great new places by building vibrant, enduring neighborhoods and communities that people, especially young people, do not want to leave.[2]

By growing and revitalizing historic town centers and ensuring that new growth and development reinforce traditional patterns, rural communities can protect the way of life that their residents treasure while supporting economic growth and bringing new opportunities. *Essential Smart Growth Fixes for Rural Planning, Zoning, and Development Codes* can help rural communities find the right tools to put their vision into practice.

2   ICMA. *Putting Smart Growth to Work in Rural Communities.* ICMA and Smart
     Growth Network. 2010. p. 1. http://icma.org/ruralsmartgrowth.

SMART GROWTH PRINCIPLES

Since the mid-1990s, the Smart Growth Network, made up of organizations representing diverse interests, has been identifying best practices, policies, and strategies that help communities get the results they want from growth.[3] The network developed 10 smart growth principles, based on experiences of communities around the country. The principles are flexible enough to apply to all types of communities, from rural to urban.

- Mix land uses.

- Take advantage of compact design.

- Create a range of housing opportunities and choices.

- Create walkable communities.

- Foster distinctive, attractive communities with a strong sense of place.

- Preserve open space, farmland, natural beauty, and critical environmental areas.

- Strengthen and direct development toward existing communities.

- Provide a variety of transportation options.

- Make development decisions predictable, fair, and cost-effective.

- Encourage community and stakeholder collaboration in development decisions.

3   For more information about the Smart Growth Network, see: Smart Growth
     Online. Smart Growth Network. http://www.smartgrowth.org/network.php.
     Accessed December 21, 2011.

RURAL COMMUNITY CATEGORIES

There are many ways to describe rural communities based on their economic, geographic, or design characteristics. Certainly, each community is unique, and rural communities can include a number of complex and contradictory qualities. However, characterizing them can help identify common challenges they might be facing as well as opportunities that could help them adopt a more sustainable approach to growth and development in the future. Most rural communities can be grouped into one of five categories,[4] though many may fall into more than one:

- *Gateway communities* are adjacent to high-amenity recreational areas such as national parks, national forests, and coastlines. They provide food, lodging, and associated services. Increasingly popular places to live, work, and play, gateway communities often struggle with strains on infrastructure and the natural environment. Many of these communities also experience seasonal population cycles that can strain resources.

- *Resource-dependent communities* are often home to single industries, such as farming or mining, so their fortunes rise and fall with the market value of that resource. A key challenge facing resource-dependent communities is diversifying the economy while maintaining their rural quality of life and character.

- *Edge communities* are located at the fringe of metropolitan areas and typically connected to them by state and interstate highways. Residents have access to economic opportunities, jobs, and services. More affordable housing and access to metropolitan amenities have made many of these edge areas grow at a faster pace than their metropolitan areas as a whole. But precisely because they are such attractive places to settle, edge communities often face pressure to continue to provide more housing and services to new residents.

- *Traditional Main Street communities* have a central commercial street as the focus of the town, with adjacent, compact, established neighborhoods. In addition, historically significant architecture and public spaces provide valuable resources upon which to build. Still, these communities often struggle to compete for tenants and customers with office parks, regional malls, and large stores that rarely locate on rural Main Streets.

- *Second-home and retirement communities* might overlap with some of the above groups, particularly edge communities and traditional Main Street communities. Like gateway communities, second-home and retirement communities struggle to keep pace with new growth while maintaining the quality of life that drew residents in the first place.

The fixes described in this publication are intended to be applicable in each of these rural community types.

HOW TO USE THIS PUBLICATION

This publication sets forth several actions that small-town and rural jurisdictions could take to address some of their most challenging growth issues. Rural communities around the country have used these actions to guide development. These essential fixes, identified by a national panel of rural smart growth experts, can address specific development issues or become a foundation for more comprehensive revisions. This publication describes policy options and does not present a recipe or a prescribed order for implementing these policies. Each community must determine what is appropriate for its needs and context.

Each essential fix contains six sections:

- **Introduction:** A discussion of the issues and growth-related challenges.

- **Response to the problem:** An overview of how local governments might respond.

- **Expected benefits:** How local governments and communities might benefit from addressing the issues.

- **Steps to implementation:** Modest adjustments, major modifications, and wholesale changes that local governments could make to their land use plans and codes to address the issues.

- **Practice pointers:** Common-sense considerations in assessing alternative implementation approaches.

- **Examples and references:** A list of general references on the topic, as well as examples of local government plans and development codes.

---

4   These five typologies were developed by the authors of *Putting Smart Growth to Work in Rural Communities* through discussions with Smart Growth Network partner organizations as well as organizations outside the network.

In addition, the chapters describe some implementation strategies, financial tools, funding sources, and related policies suited to rural areas, as well as financing and local capacity issues—such as lack of resources, investment capital, and local staff capacity to drive public-private partnerships.

While this publication is divided into 10 fixes, each fix works best when done in combination with others. For that reason, chapters sometimes refer to another chapter. For example, a discussion of directing growth toward town centers is incomplete without a discussion of protecting agricultural and natural lands outside the town. To avoid duplication, each chapter keeps to a fairly narrow discussion and assumes the reader will read the rest of the publication. Also, keep in mind that rural communities have many strategies at their disposal to determine where and how growth happens; this publication looks only at land use strategies and not at the full toolbox. Not every step to implementation is going to work the same way in each community. Regional, socioeconomic, and geographic considerations affect how and whether a particular idea might be implemented locally.

DETERMINE AREAS FOR GROWTH AND FOR PRESERVATION

INTRODUCTION

Many rural towns have found they can improve their overall quality of life by determining specific areas intended for growth and those that are to be preserved. A long-term, proactive plan establishes growth priorities. Communities can then review individual development proposals with an eye toward how they connect to comprehensive planning goals. This chapter discusses this issue and ideas for addressing it.

Rural towns and counties are recognizing that they need to designate areas where growth makes the most sense. Communities find this strategy desirable for a variety of reasons:

- It allows them to provide government services and infrastructure more cost effectively.

- It makes it easier to preserve the open space, agricultural lands, and natural resource areas that are critical to rural character and rural economies.

- It lets them provide housing in a variety of types, sizes, and price ranges with access to jobs, services, shopping, schools, and places of worship.

- It reinforces community character based on historic town patterns.

- It creates predictability and guidance for private developers to match the community vision.

- It creates more energy-efficient and sustainable communities that make it easy and appealing for people to walk or bike around town. In addition to reducing air pollution from cars, walking or biking to destinations is an easy way to get more of the daily physical activity that doctors recommend.

To accomplish these goals, local governments often need to revise their land use plans, development codes, and capital improvement plans to reinforce their community's choices about where it wants development to occur. They must also identify

Community workshops, such as this one in Bluffton, South Carolina, bring residents together to determine the most appropriate locations for future growth and development.

growth areas and make them more attractive to the development community than other areas where the community does not want development. This section focuses on strategies for growth areas and town centers.

While this chapter covers steps communities can take to identify designated areas for growth, it does not comprehensively discuss resources and ideas for supporting thriving towns and villages. A discussion of policies that relate to this topic can be found in Chapter 2 of *Putting Smart Growth to Work in Rural Communities.*[5]

RESPONSE TO THE PROBLEM

To designate growth areas in rural towns and counties, communities should undertake comprehensive planning using a participatory stakeholder and citizen engagement process. They also need analysis and data that justify the designation of specific growth areas. Justification might include fiscal impact analysis, cost of infrastructure studies, traffic modeling, water quality assessments, delineation of natural and cultural resource areas, and identification of prime agricultural lands.

---

5   ICMA op. cit., p. 17.

Many communities have used regional scenario planning, which engages participants in envisioning alternative futures and then models the impacts and benefits of several options. In this process, the resulting preferred vision is often adopted into local and regional plans and policies. The vision also typically describes what makes the community a distinctive and attractive place. Many communities use scenario planning to identify areas for preservation and areas designated for growth. The growth areas are linked by transportation networks that include roads, transit, and walking and biking trails. The preferred growth areas also typically take advantage of existing or planned improvements to other infrastructure. Although scenario planning is especially effective in high-growth areas, it can also be useful in slow-growth or no-growth environments, where growth in outlying areas can leave behind existing homes, neighborhoods, and underused infrastructure. Communities can typically conserve fiscal resources by encouraging development in areas with existing infrastructure or even in areas where infrastructure needs to be updated. However, replacing inadequate infrastructure might not always be cost-effective.

Town centers contain a concentration of land uses, usually commercial, but in many cases, residential and institutional as well. A town center can be the geographic center of a town, or a development built to serve market demand for specific land uses. If sited based on a planning and analysis process as described above, new town centers can provide a high quality of life, housing and transportation choices affordable for people with a range of incomes, many opportunities for social interaction, and cost-effective infrastructure and services. Rather than competing with existing towns, new town centers can develop a symbiotic relationship with surrounding communities through strong transportation connections, including efficient transit service where appropriate, and a shared sense of purpose created through a planning and visioning process.

Growth in many rural towns is so gradual that it is not always perceived as a concern, but at some point, some communities find that many residents oppose growth as increased development and traffic change the community's character. A clear set of principles developed through a broad community process and incorporated into the comprehensive plan can provide a framework for determining whether proposed developments fit with the desired community character and help achieve the community's economic, environmental, and social goals. The comprehensive plan and codes could also require that

Photo courtesy of Lancaster County Planning Commission

Central Market in Lancaster, Pennsylvania, is the oldest publicly owned, continuously operated market in the country. It is in the heart of an infill area that took advantage of existing infrastructure to build new offices, stores, and homes.

large development proposals include a charrette[6] to incorporate community input into their designs. For the sake of coordination and resource leveraging, it is helpful for towns to collaborate with surrounding communities to develop a regional approach to resource preservation and stormwater management and provide region-wide standards for streets that help manage stormwater runoff and are safe and appealing to pedestrians, bicyclists, drivers, and transit users.

Since a lack of in-house planning capacity can be an obstacle for small towns and rural counties, regional and state agencies often help localities find the resources to carry out these studies, support and participate in the stakeholder process, and build support for implementation. Some resources are available to enhance local capacity to pursue these strategies (e.g., economic development agency district planning funds or transportation

---

6   A charrette is a collaborative, multiday workshop that brings together stakeholders in a community to give input on a design issue or a specific development project. It allows meaningful input from the public and gives stakeholders a chance to see and react to how designers incorporate their ideas into the proposed design.

planning funds available through state departments of transportation or regional planning agencies) and to seed desirable investment and development activity.

EXPECTED BENEFITS

- The community develops a vision that values rural character and regulations and design standards to realize the vision.

- Development proposals in towns and town influence areas[7] that meet community growth goals have a more predictable review process.

- When development proposals are coordinated with community growth goals, meet local development regulations, and engage meaningful public input through charrettes, approval is usually quicker and more predictable, and the proposal generates less public opposition.

- Communities make efficient use of existing infrastructure when directing growth to designated areas. Vacant property reclamation strategies and incentives can also be key components of encouraging growth in town centers.

- Directing development to towns or town influence areas reduces pressure to develop on sensitive habitat, agricultural lands, and other open space.

- A more environmentally and economically sustainable community uses less energy by reusing existing structures and offering transportation choices, such as walking and bicycling, that can reduce greenhouse gas emissions and other pollution.

STEPS TO IMPLEMENTATION

1. Modest Adjustments

- Coordinate with nearby towns and villages to share resources, exchange ideas, and forge partnerships to build and access planning capacity.

- Identify federal grants that can be used to encourage infill and reuse of existing structures in preferred growth areas, such as the U.S. Department of Housing and Urban Development (HUD) Community Development Block Grant Program,[8] the U.S. Department of Agriculture

(USDA) Community Facilities Grant Program,[9] and the U.S. Environmental Protection Agency (EPA) Brownfields Area-Wide Planning Pilot Program.[10]

2. Major Modifications

- Identify and map the community's preferred growth areas in comprehensive plans to make it clear to developers and residents where the community wants growth to occur and to protect sensitive natural areas and prime agricultural areas.

- Establish capital improvement plans and adopt capital spending strategies—for transportation (including walking and biking facilities, public transit, and roads), public works and infrastructure, clean water programs, energy facilities, schools, and parks—that support the comprehensive plan's preferred growth areas.

- For communities that have impact or similar fees, create an incentive to develop in areas that have infrastructure to support new development by lowering the fee for those places, or encourage redevelopment of a site by using the impact fee to maintain or improve existing infrastructure. In areas with little or no infrastructure, the costs of providing and maintaining new infrastructure to support new development can be high. Factoring such costs into impact fees should be considered.

- Conduct scenario planning to identify the best areas to preserve and the most appropriate lands to develop, with modeling to measure the performance and impacts of each scenario. Use the results to inform the development of comprehensive plans and investment strategies.

- Establish community service areas that are coordinated with capital improvement plans, investment strategies, and economic development targets. Phase development with the availability of infrastructure as it is approved and constructed.

- Adopt a policy to locate all major local governmental services and offices in the town center or designated growth areas to take advantage of existing infrastructure, support the community's vision for these areas, and encourage private investment nearby.

---

7   Town influence areas are areas around a town where the town can reasonably expect to have influence over land use and planning.

8   HUD. Community Development Block Grant Program. http://portal.hud.gov/ hudportal/HUD?src=/program_offices/comm_planning/communitydevelopment/

programs. Accessed August 15, 2011.

9   USDA. Rural Development Housing & Community Facilities Programs. http://www. rurdev.usda.gov/rhs/cf/brief_cp_grant.htm. Accessed August 15, 2011.

10   EPA. Brownfields Area-Wide Planning Pilot Program. http://www.epa.gov/ brownfields/areawide_grants.htm. Accessed August 15, 2011.

3. Wholesale Changes

- Create a special expedited or prioritized review procedure to process development proposals in designated town centers. Establish development standards, such as use requirements, in neighborhood development regulations or a unified development ordinance, which is an ordinance that encapsulates zoning, subdivision standards, urban design, signage, landscaping, and other development standards that are typically separate documents.

- Designate areas for town centers in comprehensive plans where needed. Require a full range of housing types, services, and employment opportunities, and require that the new town be linked to existing development with transportation networks that accommodate public transit, walking, biking, and driving.

- Adopt an adequate public facilities ordinance (where permitted by state code) that sets criteria for utility expansion and service to outlying developments. Require that infrastructure, such as roads, water and sewer service, and schools, be in place when new development is constructed.

PRACTICE POINTERS

- Adopt a comprehensive land use map that depicts preferred development areas and describes clearly the mix of uses desired, community design principles, and the key features desired for each area.

- Town, county, and regional planning staff or municipal boards can review existing policies and determine the need to update current land use codes or undertake wholesale code revisions.

- Coordinate regionally with other local governments to adopt supportive plans and designated growth areas. Incorporate a communication and outreach plan that explains to community members how supportive plans can be implemented, what tools are available to support it (such as Economic Development Administration planning funds and state and federal transportation planning funds), and what benefits can accrue to all communities in the region.

- In many rural communities, plans, codes, and policies are often stand-alone documents, rather than fully coordinated and based on the same fundamental principles. Community staff and officials can create a process for reviewing,

coordinating, and combining these documents or at least mark reference points to illustrate connections. These efforts will help rural towns get the environmentally and economically sustainable growth they want.

EXAMPLES AND REFERENCES

Commonwealth of Massachusetts. *Smart Growth/Smart Energy Toolkit*. http://www.mass.gov/envir/smart_growth_toolkit/pages/SG-bylaws.html. Accessed April 15, 2010.

Duerksen, C. and Van Hemert, J. *True West: Authentic Development Patterns for Small Towns and Rural Areas*. American Planning Association. 2003.

Lancaster County, Pennsylvania. "Smart Growth Toolbox: Designated Growth Areas." http://www.co.lancaster.pa.us/toolbox/cwp/view.asp?a=3&q=617074. Accessed January 7, 2010.

Melious, J. *Land Banking Revisited*. Lincoln Land Institute: Cambridge, MA. 1986. pp. 20-27. http://www.lincolninst.edu/pubs/PubDetail.aspx?pubid=21.

Metro Regional Government (Oregon). "Urban Growth Boundary." http://www.metro-region.org/index.cfm/go/by.web/id/277. Accessed January 7, 2010.

Morris, M., General Editor. *Smart Codes: Model Land-Development Regulations*. American Planning Association: Chicago. 2009. http://www.planning.org/research/smartgrowth.

Nolon, J. *Well-Grounded: Using Local Government Authority to Achieve Smart Growth*. Environmental Law Institute. 2001. pp. 25-28.

Porter, D. *Managing Growth in America's Communities*. Island Press: Washington, DC. 2007. "Chapter 3: Where to Grow" and "Chapter 4: Where Not to Grow."

State of Maryland Department of Planning. "1997 Priority Funding Areas Act." 1997. http://planning.maryland.gov/OurWork/1997PFAAct.shtml.

St. Lucie County, Florida. *Towns, Villages, and Countryside* (Master Plan). 2008. http://www.spikowski.com/StLucieLDRrevisions-Ordinance06-017-AsAdopted.pdf.

Westminster, Colorado. *Design Guidelines for Traditional Mixed-Use Neighborhood Developments*. Approved May 2006. http://www.ci.westminster.co.us/Portals/0/Repository/Documents/CityGovernment/tmund.pdf.

INCORPORATE FISCAL IMPACT ANALYSIS IN DEVELOPMENT REVIEWS

INTRODUCTION

Many rural towns and counties approve developments incrementally, one project at a time, because planning for development can be hard to predict. In doing so, communities focus on short-term results, not on the long-term implications and impacts of development in aggregate. One result can be a lack of focus on long-term costs and benefits to the local government and the community as a whole. Often, they rely on rough estimates of property and other tax revenues to conclude that the proposed project will benefit the community without examining possible costs. Long-term costs include infrastructure construction and maintenance, special transit service for elderly or disabled persons, emergency services, schools and other civic facilities, and services for employees and residents of new development (e.g., affordable housing for resort workers). Failure to consider such costs before infrastructure funds have been committed can have fiscal and other impacts on residents for years through increased taxes and fewer services.

The economic, social, and environmental impacts of development are often significant. Inserting these considerations into development decision-making can help towns and counties get a fuller picture of the benefits and costs. Perhaps the most significant element for rural communities to consider is the fiscal impact of development. As many rural communities' capacities are stretched, each new development can be a relatively significant impact upon their fiscal sustainability and their ability to serve their residents. Focusing on the fiscal impact of development can help communities determine how best to allocate their resources and make development decisions that benefit residents.

The cost of the public services new residents will require and the revenues generated from new development are important to assess the fiscal impact of a project, such as the Wellington neighborhood in Breckenridge, Colorado.

RESPONSE TO THE PROBLEM

Some rural towns and counties are taking the initial step of requiring at least a basic fiscal impact analysis for all major developments. Others are going a step further by requiring that:

- The developer provide funds for a consultant (hired by the local government) who can assist the town or county in an unbiased review of the fiscal impact analysis.

- Any deficit must be addressed with funding or other mitigation measures (e.g., by donating land for a school or paying for off-site road improvements).

A simple four-step fiscal impact analysis examines the costs and benefits associated with a project:

1. Estimate the population generated by the development (e.g., the number of new residents, school-age children, and employees).

2. Translate this population into public service costs (e.g., roads, schools, and emergency services) based on costs used in the local or regional market.

3. Project the tax and other local revenues generated by the growth.

4. Compare the development-induced costs to projected revenues and, if a gap exists, determine how to address the shortfall.

While the basic methodology is straightforward, it can also include variables to compare alternative development scenarios, but only if the impact analysis is performed at a conceptual design stage. Variables could include more compact development, larger or smaller lots, adding a trail system, or deleting a school if the development shifts to senior housing (which might increase health care or emergency services costs). The analysis can also look at projected costs per phase, along with total build-out costs, so that infrastructure and expanded services can be provided in line with the estimated completion of each phase.

Where allowed by state law, concurrency regulations let the local government require that all needed infrastructure be funded and in place by the time each phase of a development is completed. If a fiscal analysis shows a development is not financially viable, the local government might choose not to approve the development. Where concurrency regulations are used, communities should consider coordinating with other municipalities in the region to ensure that development does not get pushed to locations outside of areas governed by concurrency requirements.

Once the costs of a proposed development are fully understood and communicated to the community, the local government can require mitigation measures to offset the costs. The municipality could ask the developer to propose mitigation measures to make sure the development pays its own way or to offer compensating benefits to offset community costs. Examples of mitigation measures include building a fire station, building a road connecting the proposed development to existing land uses, donating land for a school, or providing a revenue stream to pay for services the development needs. Even if local governments are not allowed to recover costs, they can still use fiscal impact analyses to help policy-makers understand the development costs and impacts and assess whether certain development types should be encouraged or discouraged in their policies and codes.

A community can conduct a fiscal impact analysis as part of a community or regional scenario planning process, rather than just in reviewing development proposals. In scenario planning, comparative costs, environmental impacts, travel choices, and other factors are used to rate the benefits and impacts of different types and locations of development across the region. Typically, more compact, mixed-use development costs less, has a lower environmental impact, and offers more transportation and housing choices.

EXPECTED BENEFITS

- Local governments will understand the full range of costs and benefits associated with a proposed development and, where allowed by state law, can ensure that costs related to infrastructure and services are recovered as part of the approval process or that mitigation is provided.

- Developments that bring demonstrated benefits to a community are more likely to attract resident and stakeholder support.

STEPS TO IMPLEMENTATION

1. Modest Adjustments

- Adopt a requirement for a full fiscal impact analysis for all major projects.

- Maintain adequate and current information on the costs of government services so that basic information for fiscal impact analyses is readily available.

- Train local government staff and planning and utilities boards to understand fiscal analysis and how it relates to infrastructure provision associated with development decisions.

- Keep capital improvement plans current and include appropriate development projections.

2. Major Modifications

- Incorporate fiscal impact analysis into county and regional scenario planning and visioning to inform the review and selection of preferred development locations.

- Identify fiscal impact thresholds that a development must meet, such as the maximum increase in bonded indebtedness or amount of remaining water or sewer capacity the community is willing to allocate to one development.

- Require fiscal impact analysis of effects on other service providers (e.g., fire districts or school districts) and surrounding jurisdictions to help ensure that neighboring communities are not burdened by the costs of providing services. If the analysis identifies adverse impacts on other jurisdictions, adopt measures to ensure mitigation (e.g., developer contributions or revenue sharing).

- Require applicants to fund adequate staff time or consulting support (with the consultant hired by the locality, not the applicant) to develop and analyze a fiscal impact assessment.

3. Wholesale Changes

- Adopt a process for measuring the long-term fiscal impacts of development. This process should consider the costs and infrastructure demands that new residents and employees will need (e.g., social services or affordable housing for lower-income workers).

- Mandate a fiscal impact analysis as part of a larger community impact analysis, including environmental, social, and economic development impacts.

PRACTICE POINTERS

- Fiscal impact analysis is an art, not a science. It requires many different assumptions about how a community will grow over time, the pace of absorption of new units in a development, changes in property tax values, and so forth. Communities should revisit impact analyses periodically to ensure that they are on target.

- Fiscal impacts vary with the type of development, its location, the level of community services it needs, and the

Photo courtesy of Albemarle Downtown Development Corporation

Schools such as this one in Albemarle, North Carolina, are community assets that can anchor neighborhoods and provide civic space and amenities for the entire community. However, the costs of adding new schools or expanding existing ones need to be considered in fiscal impact analyses.

existing capacity of services and infrastructure. The results of a fiscal impact analysis in a community with existing capacity to provide services and infrastructure will be very different from one that must build new facilities or extend existing service long distances.

- Development impacts are cumulative. One development might have minor impacts, but multiple developments over time could have significant impacts.

- A development could have a positive fiscal impact but also negative environmental and social impacts that need to be evaluated separately.

- Most residential development imposes costs on the community, which can increase over time as systems age and families have more children to enroll in school. Any developer contributions or impact fees should be used to cover anticipated costs over time instead of used once for short-term projects.

EXAMPLES AND REFERENCES

Edwards, M. *Community Guide to Development Impact Analysis.* University of Wisconsin. http://www.lic.wisc.edu/ shapingdane/facilitation/all_resources/impacts/analysis_fiscal. htm. Accessed January 8, 2010.

Florida Atlantic University, Center for Urban and Environmental Solutions. "Fiscal Analysis and Financing Tools: Fiscal Impact Analysis." *Florida Planning Toolbox.* http://www.cues.fau. edu/toolbox/subchapter.asp?SubchapterID=95&ChapterID=8. Accessed January 8, 2010.

Harrison, T. and French, C. "An Introduction to Fiscal Impact Analysis." University of New Hampshire Cooperative Extension. 2007. http://extension.unh.edu/commdev/Pubs/FIA.pdf.

Seigel, M. *Development and Dollars: An Introduction to Fiscal Impact Analysis in Land Use Planning.* Natural Resources Defense Council. 2000. http://www.nrdc.org/cities/smartGrowth/ dd/ddinx.asp.

REFORM RURAL PLANNED UNIT DEVELOPMENTS

INTRODUCTION

Local zoning codes in many areas permit negotiated developments, which are usually called Planned Unit Developments (PUDs) and can also include larger developments often called master-planned communities (MPCs). PUDs allow communities to overcome some of the strictures of conventional zoning and provide a vehicle for local government officials to negotiate community benefits, such as requiring additional open space, recreational facilities, better design, and developer contributions to infrastructure.

PUDs are often used for large areas that are master-planned by single or multiple property owners or developers. PUDs typically allow greater flexibility in layout, design, and land use than existing zoning and subdivision regulations. However, once a PUD process becomes the primary method of site plan review and permitting, municipalities sometimes are less able to connect the results of these PUDs to local comprehensive plan objectives.

Although originally intended primarily as a tool for major developments in cities and suburbs, PUDs have spread to rural areas because the process is attractive to developers, offering a more flexible way to secure approval for large developments than seeking multiple amendments and variances to a zoning code. However, the PUD approach has proliferated to the point that it has given rise to a host of unanticipated challenges. Few rural jurisdictions have the necessary staff to negotiate development agreements for complex projects. Rural development codes typically have barebones standards and processes governing PUDs and therefore provide little guidance to local officials and few controls to ensure the PUDs are properly located, are designed well, provide adequate infrastructure and community benefits, or are linked to the rest of the community.

Prospect New Town in Longmont, Colorado, is a planned unit development that used flexible development requirements to create a range of housing types and building design. Residents enjoy sidewalks, open space, and nearby services.

Rural communities are recognizing some downsides to relying on PUDs and negotiated developments:

- Large rural PUDs and MPCs often intrude and have adverse impacts on agricultural operations and natural resources, and they can strain local government services and budgets.

- Overreliance on PUDs can create uncertainty for developers when the PUDs are not tied to clear community standards to guide the development approval process. They can also create unpredictability for neighbors of proposed PUDs, who cannot rely on existing zoning or land use plans to protect their rural lifestyle.

- Environmental and design standards are sometimes overridden or ignored in the PUD review process.

- Extra work is created for staff and planning boards who have to deal with multiple mini-zoning codes created for each PUD over time. Exceptions from development standards and other requirements created for one PUD often differ from those requested by other PUDs, making consistency in decision-making difficult or impossible.

- PUDs tend to be reactive—responding to a proposed development—rather than implementing a broad, collective vision created by the community through a comprehensive plan.

RESPONSE TO THE PROBLEM

Some rural towns and counties have responded by restricting PUDs to the comprehensive plan's designated preferred development areas, forbidding the waiver of environmental and design standards, adopting updated design standards, and specifying minimum levels of community benefits such as open space and street connectivity. In other cases, towns have simply eliminated PUDs and built the necessary flexibility into their zoning codes using performance standards.

Rather than just respond to PUD proposals, small towns and rural counties can adopt zoning and subdivision provisions allowing new village-scaled development with zoning and/ or development incentives in locations where the community has decided it makes sense to grow. By mapping the areas the community wants to preserve as working lands or natural resource areas, along with areas where future infrastructure expansion would be cost-effective, a community can steer development to areas where it makes sense to build—and away from the lands it wants to preserve. Instead of waiting to react to each PUD, a community could define the type of development it wants more clearly by adopting a unified development ordinance that combines subdivision and zoning ordinances with street design guidelines, utility requirements, and open space guidelines.

Many communities have found ways to use PUDs to get development that fits with their comprehensive plan, maintains their rural character, and helps meet their overall environmental and fiscal objectives. PUDs are flexible enough to allow an attractive and environmentally sustainable design, but they often need guidelines on how to create traditional mixed-use neighborhoods. These guidelines could include subdivision, streetscape, site planning, and building design guidelines that aim to create a more pleasant, appealing, environmentally responsible, and healthy community.

For instance, a community could maintain some control over PUD applications and overall design by requiring certain

features as part of every PUD approval process. These requirements could include:

- Protection of sensitive habitat, cultural resources, and connected, usable open space.

- Street design and connectivity requirements.

- Variety of lot sizes and home sizes.

- A well-integrated mix of uses.

- Design guidelines covering site planning and general building form.

- Provisions for shared parking and on-street parking to use land efficiently.

EXPECTED BENEFITS

- Small towns and counties can use PUDs in areas where development pressures are great and where codes are not yet in place to direct growth. The PUD can provide the flexibility to establish more efficient, connected patterns with compact, mixed-use development and more cost-effective infrastructure.

- PUDs can provide increased predictability in the development review process, with a quicker, more efficient review process and less staff effort to administer development approvals.

- When certain features are part of every approval process, PUD review can require development to adhere to the community's vision and goals as established in comprehensive plans, including preserving rural character and preventing fragmentation of productive agricultural areas and environmentally sensitive and scenic natural resource areas.

STEPS TO IMPLEMENTATION

1. Modest Adjustments

- Require a mechanism, such as a charrette, to get meaningful public input starting early in the PUD review process and continuing throughout the process.

- Require applicants to pay for additional staff or consultants to help evaluate PUDs, typically through project review fees based on demonstrated costs (where allowed by state code).

- Map important natural areas and cultural resources for the town, county, or region so that as development is proposed, the PUD review process can consider these assets. This mapping will also make it easier to protect these natural and cultural resources (see Chapter 9: Protect Agricultural and Sensitive Natural Areas).

- Limit zoning and subdivision standards (especially environmental and design standards) that can be waived or modified in a PUD process, but encourage desirable development through zoning-related incentives, such as expedited permitting or priority in bonding support or other financial incentives.

- In place of PUDs, create flexible, by-right,[11] mixed-use zone districts adjacent to towns and in town influence areas to accommodate large developments that are in accord with town or county comprehensive plans.

2. Major Modifications

- Establish a minimum list of public benefits that the applicant must commit to providing prior to PUD approval (e.g., setting aside a certain percentage of the site as permanently protected open space).

- Require all PUDs and MPCs to be in accord with comprehensive plan requirements, particularly locating in the plan's preferred growth areas.

- Encourage mixed-use zoning in PUDs, including commercial development that fits the scale of the community, reinforces a sense of place, and promotes walking or biking, such as small stores, community centers, or offices.

- Require a fiscal impact analysis for the PUD process and require that the PUD demonstrate a long-term fiscal benefit to the community.

3. Wholesale Changes

- Require evaluation of PUDs based on street connectivity, lot and home size variety, integration of a mix of uses, adherence to design guidelines, open space connectivity, and parking strategies.

- Create a set of neighborhood development types (high-, medium-, and low-density as well as mixed-use) with related design guidelines that can be the basis for PUDs, and adopt these types into zoning codes. This will help avoid lengthy approval periods, excessive review time, and poor locations.

- Prohibit the use of PUDs in all rural and agricultural zone districts outside of town influence areas unless they are in an approved new town location.

- Strengthen PUD requirements to promote environmental and design standards.

PRACTICE POINTERS

- Consider establishing a detailed list of community benefits expected in return for variations to the desired uses, design, and locations that the community has established. Benefits might include a specified amount of permanently preserved open space, reclamation of degraded sensitive areas, or improvements to roads and other infrastructure. The list provides reassurance to the community and some predictability for developers.

- Give priority to PUD or MPC applications that are in the town, adjacent to the town, or in town influence areas, with additional preference to proposed developments that incorporate existing structures or redevelop on vacant properties.

- To the maximum extent possible, use development standards from existing zoning and subdivision ordinances to avoid creating PUDs that are mini zoning districts and difficult to administer.

---

9   "By-right" means that the project is permitted under current zoning and needs no special review or approval.

EXAMPLES AND REFERENCES

Benton County, Oregon. *Benton County Development Code*. "Chapter 100: Planned Unit Development in Corvallis Urban Fringe." Adopted 1990. http://www.co.benton.or.us/cd/planning/documents/dc-ch_100.pdf.

Center for Land Use Education. "Planning Implementation Tools: Planned Unit Development." University of Wisconsin-Stevens Point. 2005. http://www.uwsp.edu/cnr/landcenter/pdffiles/implementation/PUD.pdf.

City of Mount Vernon, Washington. *Planned Unit Developments: Handbook and Site Planning Guide*. 2006. http://www.ci.mount-vernon.wa.us/imageuploads/Media-1064.pdf.

McMaster, M. "Planned Unit Developments." *Planners Web*. 1994. http://www.plannersweb.com/wfiles/w490.html.

Northwest Vermont Project. "Transportation and Land Use Connections: Planned Unit Development." http://www.transportation-landuse.org/pages/tools/pud.htm. Accessed January 8, 2010.

New York State Legislative Commission on Rural Resources. *A Guide to Planned Unit Development*. 2005. http://www.dos.state.ny.us/lg/publications/Planned_Unit_Development_Guide.pdf.

St. Lucie County, Florida. *Towns, Villages, and Countryside* (Master Plan). 2008. http://www.spikowski.com/StLucieLDRrevisions-Ordinance06-017-AsAdopted.pdf.

USE WASTEWATER INFRASTRUCTURE PRACTICES THAT MEET DEVELOPMENT GOALS

INTRODUCTION

Finding wastewater management solutions for new developments, revitalizing areas, and failing systems is critical to protecting water quality and human health. Many rural towns across America want to direct growth to the most suitable areas, such as near fire stations and schools, or extend existing villages, but they are struggling to find the most appropriate wastewater infrastructure solution, and some approaches can have unintended consequences.

Additionally, many rural communities and small towns must address failing wastewater systems, including septic systems. Addressing the environmental and public health concerns associated with failing septic systems can be difficult in small towns and rural areas. Management, maintenance, and compliance can be challenging, particularly in smaller communities, for all types of wastewater treatment. This issue is particularly relevant in states that are largely rural or have not widely installed sewer service. Seventy percent of Vermont towns, for example, do not have public wastewater treatment.[12] Communities without sewers tend to be small. In Indiana, for example, 88 percent of communities without sewers have 200 or fewer homes; in Iowa, incorporated communities without sewers have 64 homes on average.[13] Based on the size and location of these communities, it is often not feasible to extend to them sewer lines from existing treatment plants.[14]

Selecting the appropriate wastewater management system can help communities protect their water resources. The city of Bayfield, Wisconsin, on the shore of Lake Superior, worked with the surrounding township to build a regional wastewater treatment plant that would better protect the lake and help preserve the community character and clean water that attract tourists.

Photo courtesy of James Bilbrey via Flickr.com

The design and location of a community's wastewater infrastructure can affect its future development patterns, natural and agricultural areas, and health of watersheds.

RESPONSE TO THE PROBLEM

Rural communities and small towns come in all shapes and sizes, as do their corresponding wastewater infrastructure needs and solutions. No single solution will be appropriate for every community. Understanding the relationship between wastewater infrastructure and community growth can help communities make better choices and protect water quality, human health, and the environment.

An important first step for any rural community is to protect existing investments, which includes identifying what systems are currently in place and their state of repair. Rural communities and small towns can inventory existing systems, educate households with septic systems about the importance of regular system maintenance, and require all systems in their jurisdiction to be inspected and maintained. When poorly managed and

---

12 Vermont Department of Housing and Community Affairs. "Background Report: Improving Wastewater Treatment Options for Vermont's Unsewered Villages." 2006. http://www.dhca.state.vt.us/Planning/VillageWastewater.htm.

13 Cunningham, S. L. *Do You Want Utilities With That? Avoiding the Unintended Economic Consequences of Poorly Planned Growth on the Provision of Water and Sewer Service.* Center for Environmental Policy and Management, University of Louisville. Summer 2006. http://cepm.louisville.edu/Pubs_WPapers/practiceguides/PG14.pdf.

14 EPA. *Handbook for Managing Onsite and Clustered (Decentralized) Wastewater Treatment Systems: An Introduction to Management Tools and Information for Implementing EPA's Management Guidelines.* 2005; updated 2010. http://cfpub.epa.gov/owm/septic/septic.cfm?page_id=289.

maintained systems fail to adequately treat wastewater, the municipality can end up bearing the cost of upgrading the systems.

Rural communities and small towns can reap significant savings by investing in their existing water infrastructure. In tough economic times, regular maintenance expenditures can become targets for budget cuts, especially when the infrastructure is underground and only "seen" when problems arise, such as sewage flows into nearby streams. But the costs of repairing problems, including degraded streams and possible loss of tourist revenue, can be higher than the costs of regular maintenance.

Planning for growth is essential for rural communities that want the benefits associated with growth while preserving their rural character. When development design and open space preservation are decided one subdivision at a time, rural communities can lose their ability to take advantage of excess capacity or leverage a planned wastewater system to accommodate nearby growth. Focusing on individual lots or even individual neighborhoods forces the community to address wastewater needs site by site, which can be ineffective at protecting water quality or supporting growth. Processes like visioning exercises (see Chapter 1: Determine Areas for Growth and for Preservation) can help communities choose the type and location of development they want.

In addition, rural communities could consider regional planning goals in addition to their own goals for growth and development. Looking at the broader region also allows communities to consider cumulative impacts on the watershed from their development decisions and to leverage and coordinate their wastewater infrastructure strategies and investments. Then communities can choose a wastewater management system that is consistent with their vision for growth, supports that growth, and protects public health and the environment.

Several types of wastewater systems are available to rural communities and small towns. Not all of these systems are appropriate for all types of rural communities, as some systems can contribute to dispersed development patterns, ineffective natural resource protection, and fiscal inefficiencies. By selecting and using appropriate wastewater infrastructure, rural communities can protect their water quality and public health in a way that supports their other community goals, such as maintaining rural character or promoting thriving town centers. Wastewater system options include:

- *Septic systems.*[15] Rural communities are often served by conventional on-site septic systems, as they work well for single homes in remote areas. However, traditional septic systems might not be appropriate to support a new subdivision or cluster of new homes. Using individual septic systems in these scenarios without corresponding development planning can encourage low-density, dispersed development, which can significantly alter the rural landscape and degrade natural resources.

- *Cluster systems.*[16] Cluster systems can create more compact development and can help support a rural community's growth goals. However, using these systems outside of a comprehensive development plan can lead to the creation of tiny pockets of housing that break up large, contiguous agricultural or natural areas and are far from jobs, schools, stores, or other amenities. To use these systems effectively, rural communities should use them in the areas they have designated for growth.

- *Advanced technologies.*[17] Advanced treatment technologies generally have a smaller footprint and can treat more wastewater on less land, which can allow more compact development. They also can treat wastewater in amounts comparable to centralized sewage systems, which means larger areas or neighborhoods can be serviced. However, if applied outside of the context of a comprehensive development plan, advanced technologies can allow development in areas not accessible for conventional treatment, such as areas that communities want to preserve as open space or farmland. Like cluster systems, without a comprehensive development plan, these systems can facilitate dispersed development patterns and are most effectively used in areas designated for development.

---

15 A septic system is a type of decentralized wastewater treatment that consists primarily of a septic tank and a soil absorption field. Each septic system typically occupies a relatively large area, and systems must have adequate spacing and distance from wells and surface waters.

16 A cluster system, also called a shared or community system, is a type of decentralized wastewater treatment system that serves more than a single home or business.

17 Advanced treatment systems encompass a broad range of technologies. The unifying feature is a separate treatment unit next to the septic tank that treats the effluent before it is discharged to the drainfield (a below-ground absorption field, also called a leach field).

- *Centralized sewerage.*[18] Centralized systems have typically been used in cities and towns. Over the past several decades, centralized systems have been used to expand into farmland or other rural landscapes at the edge of established communities. In addition, some communities have used centralized treatment to replace failing septic systems with the goal of protecting public health. However, expanding centralized sewer systems without a development plan can enable and encourage dispersed development in rural areas, which can create pressure to attract additional ratepayers to support a wastewater treatment plant and conveyance system. A centralized sewer system can attract development regardless of whether it is in the most appropriate area for growth. Rural communities might want to limit the expansion of centralized treatment to existing developments and established planned growth areas. Doing so also allows coordination with other investments in transportation, housing, and jobs.

One important and often overlooked strategy for communities is identifying where existing wastewater infrastructure has excess capacity. Neighborhoods with existing (or excess) capacity could support additional growth. This strategy can be effective at accommodating new development within existing system limits.

Regardless of the system used, communities might need to align local development regulations with wastewater treatment standards to support a range of wastewater systems. For example, local regulations sometimes limit the use of some types of decentralized systems, rather than requiring a certain level of performance and allowing any system that can achieve that performance level. Such regulations can lead communities to choose other systems that might not be adequate to handle the community's wastewater, which could then degrade public health and water quality or lead to an expensive sewer expansion that encourages dispersed development. In addition, codes for new on-site wastewater treatment systems should be consistent with existing and future land use plans.

Additionally, some municipalities have used wastewater treatment standards that prohibit new decentralized wastewater treatment systems as a way to rein in growth. However, such standards can have the unintended effect of restricting wastewater treatment options that are compatible with development goals. For instance, many communities have sites where development is desired or has already occurred but centralized sewerage is financially or logistically impractical. These communities need the flexibility to choose wastewater treatment options that protect water quality while allowing growth and development.

A pressing problem for many rural communities is how to address failing septic systems, which pollute groundwater and cause water quality problems for nearby water bodies. A common response to this problem is to replace these systems with centralized wastewater treatment, which can lead to additional growth in areas that the community would prefer to remain undeveloped and create pressure to operate and maintain sometimes complex centralized systems. Many times, addressing these failing septic systems is a priority for the local and state government, but the challenge is to how to address the problem without inadvertently encouraging development in areas not intended for growth. Incremental approaches could include:

- Offer incentives or technical assistance to homeowners to replace their failing septic systems. In some rural areas, neighborhoods with failing septic systems are near an important natural resource, such as a lake or mountain range, which is an economic driver for the community. In these instances, the municipality might be able to leverage local businesses to help create an incentive fund.

- Create a municipal septic management district or a responsible management entity (RME)[19] responsible for the repair, replacement, and maintenance of homeowners' septic systems. In this case, the municipality or the RME can pay for or organize the replacement of the failing system. The RME would then be responsible for the ongoing maintenance. The homeowner would pay a fee for this service, similar to the sewer fee homeowners pay on centralized treatment systems. Wisconsin uses this approach.[20]

---

18 Centralized sewerage collects and transports household sewage via a network of pipes and pump stations to a municipal treatment plant. Most commonly used in cities and small towns, centralized treatment systems treat waste flows and protect water quality but are also the most expensive system.

19 For more information on RMEs, see: EPA. *Voluntary National Guidelines for Management of Onsite and Clustered (Decentralized) Wastewater Treatment Systems.* 2003. http://www.epa.gov/owm/septic/pubs/septic_guidelines.pdf.

20 Wisconsin Department of Safety and Professional Services. "Safety and Buildings Division Private Onsite Wastewater Treatment Systems." http://dsps.wi.gov/sb-powtsprogram.html. Accessed January 5, 2012.

- Create indicators or criteria to determine when a neighborhood with failing septic systems might be a good candidate to connect to a centralized system and when it should consider different alternatives. For example, areas planned for additional growth with moderate densities might be better candidates for centralized systems. Areas not planned for growth or for very low densities, such as one unit per 20 or more acres, might be better suited to septic replacement. Possible criteria for centralized systems could include:

  - Any structure served by an expansion must be on a site with access to existing roads, water, and utilities and within or contiguous to existing development.

  - Collector lines connecting a home or business to the main trunk line must be no longer than 1,000 feet.

  - Additional infrastructure investments, such as transportation, schools, or additional housing, are likely.

  - The context, including density of surrounding development, condition of surrounding wastewater systems, or proximity to an existing or emerging town center or employment center, is appropriate for a centralized system.

Considerable costs can be associated with wastewater treatment systems, especially if the new system is intended to support a new development or housing cluster. Building, operating, and maintaining new infrastructure can divert money from badly needed repairs and upgrades to existing infrastructure, so rural communities need to carefully consider where and how to pay for new wastewater infrastructure. Many different strategies are available to help rural communities maintain and finance their wastewater infrastructure, including:

- *Impact fees.* Some communities require new developments to pay an impact fee that would finance the wastewater system construction costs. As part of this strategy, communities could consider requiring long-term financial maintenance plans for any new decentralized system when reviewing plans for approval. If such a plan is not established before installation of these systems, municipalities might find themselves responsible for the continued operation, maintenance, and repair of failing systems.

- *Performance bonds.* A community could require a performance bond for any decentralized system, which could provide the community with some guarantee of the effectiveness of the installed system. A performance bond or escrow account could be used to cover future operation and maintenance costs.

- *Land tax.* The community could require any development on pristine land to pay a premium land tax. These funds could then be used to support the repair or replacement of failing systems as well as the revitalization of older neighborhoods or town centers.

- *Maintenance agreements.* Rural communities could require maintenance agreements between a property owner and a maintenance firm or the municipality. These agreements could provide the rural community some guarantee of effective management and maintenance of the new system.

There is no single, simple solution for managing wastewater in rural communities and small towns. Planning for growth and examining the range of possible consequences from water infrastructure investments is critical. Doing so allows the community to balance its water infrastructure needs, such as accommodating new growth or alleviating an existing problem, with its environmental and public health protection goals.

EXPECTED BENEFITS

- Aligning land use policy and public investments in water infrastructure can help rural communities and small towns save money by concentrating services.

- By addressing wastewater needs, rural communities can provide additional capacity for growth, which can enhance the potential for economic development. Providing attractive options for in-town development can protect the rural character of outlying areas.

- A comprehensive regional plan for wastewater treatment infrastructure can improve water quality, protect public health, safeguard investments in existing infrastructure, and ensure that land use plans can be implemented as desired.

- A user-funded management program for decentralized wastewater treatment systems can protect public health and local water resources while allowing growth in town centers.

STEPS TO IMPLEMENTATION

1. Modest Adjustments

- Establish processes to align water infrastructure investments with other public investments such as transportation, housing, and schools.

- Inventory existing wastewater infrastructure, assess current conditions, and update this inventory regularly.

- Identify excess capacity in existing infrastructure so that development can be directed to areas that can support additional growth, making the most of infrastructure investments.

- Develop "fix it first"[21] criteria for water infrastructure investments.

- Establish a public education program on the importance of regular maintenance for septic systems and support homeowners with regular inspections and technical assistance.

- Revise local regulations if necessary to allow the range of decentralized systems that are able to meet performance standards consistent with local water quality goals and land use plans.[22]

2. Major Modifications

- Delineate growth areas where compact development can be located, and create policies that direct development into those areas based on infrastructure availability and preservation of open space. Designate areas for new investments in water infrastructure. Reinforce these designations in all plans, policies, and regulations.

- Require long-term financial maintenance plans for any new water infrastructure, particularly decentralized systems, when reviewing plans for approval.

- Establish a program to manage all decentralized wastewater treatment systems, including requiring homeowners to have their systems inspected or pumped on a regular schedule and to repair or replace failing systems and cesspools as needed.

- Require users of decentralized systems, such as septic or cluster systems, to pay regular service fees to fund the regular maintenance and management of these systems, just as users of centralized sewerage facilities pay for comparable services.

- Require developments in previously undeveloped areas to finance all their wastewater system construction costs.

- Require performance bonds for new, noncentralized wastewater systems.

- Charge initial impact fees and/or assess a regular utility fee to cover county or regional management oversight costs, including the development of a tracking program to oversee maintenance and staff time spent on ensuring compliance and conducting inspections.

3. Wholesale Changes

- Establish a mechanism for regional planning of wastewater infrastructure that can cut across political boundaries and overcome fragmented system ownership and operation.

- Develop a policy for decentralized systems, particularly septic systems, that includes processes for permitting such systems, replacing failing systems, and identifying when centralized treatment might be warranted.

- Create a septic management district or responsible management entity.

PRACTICE POINTERS

- Base wastewater treatment decisions on the community's water quality, public health, and land use goals.

- Assess capacity in existing treatment plants to determine where planned growth can be accommodated.

- Price services to reflect the full cost of building, operating, and maintaining a system. Accurate pricing is critical to ensure proper and efficient operations and to send a signal to customers about the true cost of treatment options for different types of development.

---

21 Under a "fix it first" policy, a community invests in fixing and maintaining existing infrastructure (e.g., roads and bridges) before it spends money on constructing new infrastructure.

22 The National Onsite Wastewater Recycling Association has developed a Model Code Framework to help states and localities resolve conflicts with the permitting and use of decentralized systems. See: National Onsite Wastewater Recycling Association. *Model Code Framework for the Decentralized Wastewater Infrastructure.* 2007. http://www.modelcode.org/publications.html.

EXAMPLES AND REFERENCES

10,000 Friends of Pennsylvania. *Sewage Facilities and Land Development*. 2005. http://10000friends.org/sewage-facilities-and-land-development.

Doylestown Township, Pennsylvania. *Septage Management Program*. 2005. http://www.doylestownpa.org/SMP.htm.

EPA. *Decentralized Wastewater Treatment Systems: A Program Strategy*. 2005. EPA 832-R-05-002. http://www.epa.gov/owm/septic/pubs/septic_program_strategy.pdf.

EPA. *Protecting Water Resources with Smart Growth*. 2004. EPA 231-R-04-002. http://www.epa.gov/smartgrowth/water_resource.htm.

EPA. *Voluntary National Guidelines for Management of Onsite and Clustered (Decentralized) Wastewater Treatment Systems*. 2003. EPA 832-B-03-001. http://www.epa.gov/owm/septic/pubs/septic_guidelines.pdf.

Johnstone, S. et al. *Smarter Land Use with Onsite Systems: One State's Process*. 2004. http://www.stone-env.com/docs/prespaperabs/StoneWW-LandUseOnsiteMaine_paper.pdf.

Joubert, L. et al. *Creative Community Design and Wastewater Management*. Prepared for the National Decentralized Water Resources Capacity Development Project, Washington University, St. Louis, MO, by University of Rhode Island Cooperative Extension, Kingston, RI. 2004. http://www.uri.edu/ce/wq/NEMO/Publications/PDFs/WW.CreativeDesignAndManagement.pdf.

Magliaro, J. and Lovins, A. *Valuing Decentralized Wastewater Technologies: A Catalog of Benefits, Costs, and Economic Analysis Techniques*. Rocky Mountain Institute. 2004. http://www.rmi.org/rmi/Library/W04-21_ValuingDecentralizedWastewater.

National Onsite Wastewater Recycling Association. *Model Code Framework for the Decentralized Wastewater Infrastructure*. 2007. http://www.modelcode.org/publications.html.

Stone Environmental, Inc. *Decentralized Wastewater and Water Supply: Further Reading*. Prepared for the Maine State Planning Office. 2003. http://www.stone-env.com/docs/reports/StoneWW-Me09TAB12DecentReading.pdf.

Vermont Department of Housing and Community Affairs. *Wastewater Solutions for Vermont Communities*. 2008. http://www.dhca.state.vt.us/Planning/SewageSolutions/WW_SolutionsVermont.pdf.

RIGHT-SIZE RURAL ROADS

INTRODUCTION

Rural roadways help define rural character and community image—from a narrow, winding road through the mountains to a walkable, tree-lined neighborhood street to a bustling downtown Main Street. Many residents in rural areas want safe roads that also maintain a rural character and avoid the uniformity frequently imposed by conventional roadway design standards. State departments of transportation and local governments are also concerned about ever-increasing costs to extend and maintain roads required by dispersed, large-lot development. The ownership, funding, operation, and design control of streets is complex, with roads owned and operated by cities, towns, counties, state agencies, or even private entities and often subject to federal transportation policies, further complicating transportation and redevelopment efforts.

In many rural towns, the Main Street is a state road and under state control. Fast-moving through traffic comes through these towns' central business districts, which can make it difficult for the towns to maintain traditional Main Streets with local-serving stores and a strong sense of community character. As the street needs to serve not only local residents, but also freight and through traffic, redevelopment can be challenging. However, it can also be an opportunity to work with the state department of transportation to use transportation funding to redesign a road so that it works better for the community as well as for through traffic.

Communities across the country are investing in streetscape projects, area planning, and rezoning to encourage infill development along their commercial corridors. There are usually economically obsolete and/or underused real estate assets, known as greyfields, and brownfield properties along these aging corridors, often at key intersections and within walking distance of surrounding residential neighborhoods. Redevelopment on

Boyne City, Michigan, has revitalized its downtown in part by ensuring that pedestrians and bicyclists, as well as drivers, feel safe and welcome. The crosswalk and curb bulb-outs narrow the street to calm traffic and make crossing the street easier.

Photo courtesy of Harbor House Publishers

underused or vacant properties can provide housing near services and current or potential transit routes. The current or future transit service typically available along these corridors, coupled with nearby walkable destinations, offers more convenient and affordable transportation choices for residents. Because many of these corridors are state highways, communities can often combine state transportation funding with local funding and developer investments for cost-effective enhancements that improve the street's aesthetics, traffic capacity, and safety for all users.

Related non-transportation infrastructure, such as water, sewer, and stormwater systems, also faces fiscal challenges due to decades of expansion and increasing costs for maintenance and replacement. By coordinating planning and project development for these systems with transportation networks and land use, communities can use their limited funds more efficiently to develop more compact, cost-effective systems. This coordination will particularly help stormwater systems, which can be overloaded with runoff from wide streets.

RESPONSE TO THE PROBLEM

Many communities are finding new approaches to balance the needs of local pedestrians, shoppers, employees, business owners, and residents with the need for through traffic, including freight, to move safely and efficiently. Balancing these needs recognizes that good state highways and strong Main Streets are both critical to a community's economic vitality.

Narrower streets naturally calm traffic, while wider streets encourage faster driving regardless of posted speed limits. Pedestrians and bicyclists feel less safe near fast-moving traffic. In districts like Main Streets where a community wants to encourage foot traffic to support stores, pedestrians must feel safe and comfortable walking along and crossing streets. The same street design changes that calm traffic also make streets more attractive, are safer for pedestrians and bicyclists, and can help protect a historic Main Street's distinctive character. Extending walkable streets through neighborhoods gives residents more choices for getting around, and making it safe and convenient to walk or bike helps people to incorporate regular physical activity into their daily routines as recommended by the medical community. Complete streets—streets that are designed for pedestrians, bicyclists, transit users, and drivers—provide these options for residents.

A good walking environment in rural areas and around towns can include trail networks that are fully integrated with the on-street pedestrian and bicycle network, so that residents can use trails and greenways from outlying areas to get to and from town, not just for recreation. An integrated network of complete streets and trails should connect rural and in-town neighborhoods, transit routes, downtown, neighborhood parks, and recreation areas, so that walking, biking, and transit are fully supported transportation choices. The network should include safe street crossings using techniques appropriate to the town's character and context, such as mid-block crosswalks, median islands, curb bulb-outs to shorten crossing distances, or roundabouts at key intersections. A well-connected network gives people more route choices instead of forcing all traffic onto one wide arterial street, so streets can be narrower. Typically, allowing narrower streets requires adjusting the subdivision ordinance and street specifications. Making sure that streets are right-sized—in other words, only as big as required—can save on construction and operating costs.

Outside of the downtown, many rural towns have corridors of spread-out stores and other commercial uses. In many places, the streetscape is designed for cars to move quickly, not for people to walk. Redeveloping these corridors is an effective way to add new housing, shopping, and community facilities near existing neighborhoods. Communities can also improve stormwater management by using green infrastructure features, like swales, rain gardens, or pervious paving, during redevelopment for both new and rebuilt streets and parking lots. As part of the comprehensive plan and zoning updates, revisions to subdivision and street design guidelines or streetscape standards could include:

- Revisions to the road classification system to incorporate a gridded street network.

- Reduced design speeds to allow narrower streets and wider sidewalks.

- Reduced street width standards for most local and connector streets.

- Intersection designs with reduced turning radii and safe pedestrian crossings.

- Street trees in tree wells large enough to accommodate their root systems to create a continuous shade canopy and to capture, slow down, and infiltrate rainwater.

- Green infrastructure stormwater management features to promote infiltration.

- Street lights at a height that provides good lighting for pedestrians as well as drivers, with fixtures that direct the lighting to the street and preserve dark skies.[23]

- Standards ensuring pedestrian and bicyclist safety, particularly around schools.

EXPECTED BENEFITS

- Connected street networks, combined with compact development and right-sized streets, give residents and visitors more choices in how they get around, which can help reduce traffic congestion on major roads.

- Narrower streets with traffic-calming features are safer, with fewer and less serious crashes due to slower travel speeds.

---

23   For sample Dark Skies ordinances, see: International Dark Sky Association. Home Page. http://www.darksky.org. Accessed December 21, 2011.

- Narrower streets use less pavement, which can be coupled with green streets techniques—using vegetation and permeable surfaces to manage stormwater at its source, make walking and bicycling more appealing, and beautify the streetscape—to reduce runoff and improve water quality.

- More attractive and safer streetscapes help support redevelopment and economic prosperity by making these public spaces more inviting and encouraging foot traffic that brings more customers to stores.

- Transportation options, especially biking and walking, help promote healthier, active lifestyles while reducing greenhouse gases and other pollution. They can also help reduce the costs of owning and operating a vehicle.

STEPS TO IMPLEMENTATION

1. Modest Adjustments

- Conduct a walking audit of neighborhood streets, reviewing the street widths and other characteristics, including those that seem to work well, as a first step in developing new street design guidelines based on the existing characteristics.

- Start a street tree planting program, since shade and buffering from vehicles are critical to pedestrian comfort; street standards could encourage or require tree-lined streets.

- Encourage and permit rain gardens and other green infrastructure techniques to slow, filter, and absorb water while making the street greener. Rain gardens and similar techniques can provide a landscaped zone between the sidewalk and travel lanes, buffering pedestrians from the speed, noise, and danger of moving traffic, or can be installed in curb extensions at crosswalks.

- Conduct a parking survey to count all available public and private parking spaces in the downtown area as a first step in developing a parking strategy. This strategy should look realistically at the amount and location of parking needed for the entire district, rather than requiring each property to provide all of the parking spaces potentially required for its operations.

- Create a bike/pedestrian plan to identify ways to make walking and bicycling safer and more appealing.

2. Major Modifications

- Develop and adopt street connectivity regulations for new development areas.

- Try a "road diet" that reduces the number of through-lanes on a street by allocating excess capacity to parking lanes, bike lanes, landscaped medians, or sidewalks. After careful review of current and projected traffic numbers, many communities have found that four- and five-lane roadways can be reduced to two- or three-lane configurations. Some communities have found that doing a sample road diet on a few blocks of a single street creates a demonstration project that helps show the benefits and low negative impacts of narrower, greener streets.

- Encourage alleys in compact, walkable residential districts, but with a narrower paved or graveled width (usually 10 to 12 feet) and an easement for utilities (usually 20 feet overall). In a residential grid, alleys should connect across blocks to make garbage pickup easier. In commercial areas, most communities that have alleys require them to be at least 24 feet wide to allow dumpster access and deliveries.

- Require shared parking for commercial businesses, public and community facilities, and downtown developments. Develop a parking management plan to take advantage of existing supply, and reduce parking requirements for new buildings and redevelopment accordingly.

- Create and implement a comprehensive streetscape improvement plan for major commercial corridors to improve access for public transit, bicyclists, and pedestrians.

3. Wholesale Changes

- Adopt a complete streets policy to require bike, pedestrian,and transit facilities on all new or rebuilt local roads.[24]

- Conduct a planning study for a major corridor to re-engineer the roadway and plan for development that will be "transit ready" when bus or other transit comes. Communities can implement this approach gradually through site-planning requirements, modifications to mixed-use requirements,

---

24   According to the National Complete Streets Coalition, as of December 2011, 314 communities have adopted or pledged to adopt complete streets policies. For more information, see: National Complete Streets Coalition. "Complete Streets Atlas." http://www.completestreets.org/complete-streets-fundamentals/complete-streets-atlas.

density requirements, and parking regulations as the transit system is enhanced and extended.

- Convene a regional task force—including representatives from counties, towns, regional agencies, and the state department of transportation, among others—to review policies, guidelines, and underlying legislation and help determine changes that would allow and encourage new development to be more compact and connected, with less environmental impact and safer, more convenient transportation choices. In addition to interagency coordination, identify potential modifications to regional or state standards, such as street connectivity, access management, and drainage standards, that would make it easier for localities, developers, and builders to deliver more environmentally sustainable transportation networks and communities.

- Require that all new roadways and trails follow design and connectivity standards and that any new development reserve terminus points to adjacent undeveloped property for future required connection.

- Convert one-way streets to two-way streets to improve walkability and mobility and make it easier for customers to reach businesses in the town center.

Photo courtesy of Dan Burden, Walkable & Livable Communities Institute

Hamburg, New York's Main Street has on-street parking, which helps calm traffic; bike lanes marked with colored pavement; and clearly marked crosswalks with curb bulb-outs to shorten crossing distances.

PRACTICE POINTERS

- Using green streets techniques during redevelopment of commercial properties for both new and rebuilt streets and parking lots can better manage stormwater while making the street more attractive and appealing.

- Review redevelopment standards and regulations to identify obstacles, determine possible incentives, and encourage redevelopment of properties along existing roadways.

EXAMPLES AND REFERENCES

Bray, T. and Rhodes, V. "In Search of Cheap and Skinny Streets." *Places*, Vol. 11:2. 2006. pp. 33-39. http://www.cues.fau.edu/cnu/docs/In_Search_of_Cheap_and_Skinny_Streets-Bray-Rhodes.pdf.

Maryland State Highway Administration. *When Main Street is a State Highway*. 2002. http://www.marylandroads.com/ohd/MainStreet.pdf.

New York City Department of Transportation. *Street Design Manual*. Revised July 2010. http://www.nyc.gov/html/dot/html/about/streetdesignmanual.shtml.

Oregon Department of Transportation. *Main Street… When a Highway Runs through It: A Handbook for Oregon Communities*. November 1999. http://www.oregon.gov/ODOT/HWY/BIKEPED/docs/mainstreethandbook.pdf.

Pedestrian and Bicycle Information Center. "Walkability Checklist." http://www.walkinginfo.org/library/details.cfm?id=12. Accessed December 21, 2011.

Seattle Department of Transportation. *Right-of-Way Improvements Manual Version 2.0*. Revised May 2011. http://www.seattle.gov/Transportation/rowmanual.

U.S. Green Building Council. *LEED for Neighborhood Development Rating System*. Updated May 2011. http://www.usgbc.org/DisplayPage.aspx?CMSPageID=148.

Virginia Department of Transportation. "Secondary Street Acceptance Requirements." http://www.virginiadot.org/projects/ssar. Accessed April 15, 2010.

Washington State Department of Transportation. *Understanding Flexibility in Transportation Design—Washington*. April 2005. http://www.wsdot.wa.gov/Research/Reports/600/638.1.htm.

6 ENCOURAGE APPROPRIATE DENSITIES ON THE PERIPHERY

INTRODUCTION

Rural communities generally want to remain rural or maintain their small-town character. Many of these communities encourage low-density development in the belief that it will maintain the rural character. However, low-density developments are usually more suburban than rural in nature and frequently use suburban standards for streets, landscaping, setbacks, and lot sizes. For communities trying to preserve rural character, development of 2- to 10-acre lots is particularly challenging. Lots of this size pose a host of problems that often undermine rural character and make it difficult to protect natural and fiscal resources. These include:

- Infrastructure and services are more costly and inefficient to provide.[25]

- Residents demand services, such as road maintenance and recreational facilities, but the supporting tax base is inadequate to provide these services.

- Productive agricultural lands and sensitive natural areas are fragmented, which makes farming or ranching more difficult and disrupts natural habitats.

- Domestic animals and trash are introduced into agricultural areas and wildlife habitat.

- Future town-level development is often difficult or impossible if the development does not include easements for central water or sewer lines or drainage or has limited and disconnected road rights-of-way.

- These lots often rely on septic systems, which can fail (see Chapter 4: Use Wastewater Infrastructure Practices That Meet Development Goals).

Photo courtesy of EPA

Development on the edge of town, as in Bel Air, Maryland, can include walking paths to transition between homes and open space.

- Directing growth to existing towns uses infrastructure in which public money has already been invested. Development that is outside of these areas does not take full advantage of these taxpayer investments.

- Large, spread-out lots make it difficult to walk or bike to destinations, forcing residents to drive everywhere, increasing air pollution and greenhouse gas emissions from driving and making it less convenient for people to work regular physical activity into their daily routines.

The density of development helps shape the character of a community. High rises evoke big cities; subdivisions of single-family homes are typical of many suburbs. Farms, villages, and towns with small, walkable downtowns are typical of rural settings. Densities vary by place and circumstance; one key to preserving a sense of place and improving the community is to use the appropriate density for the context.

Rural communities often allow land development patterns that are not dense enough to provide cost-effective services and infrastructure, but that are too dense to maintain a truly rural feel. Such development patterns typically fragment agricultural

---

25 For example, one study describes the potential infrastructure and development cost savings of traditional neighborhood development versus conventional development. See: Ford, J. "Comparative Infrastructure & Material Analysis of Smart Growth & Conventional Projects." Morris Beacon. January 13, 2010. pp. 3-6. http://www.morrisbeacon.com/media/portfolio-projects/research/MBD-EPA-infrastructure.pdf.

lands and natural resource areas, which can harm the area's economic and environmental health.

Typical housing densities of about two to four units per acre close to town, and one unit per 2 to 10 acres in more rural areas, can create problems for rural communities. These densities result in lots that are too big to mow easily and usually too small to farm. One narrow circumstance in which this pattern can work is in some areas near cities, where 5 to 10 acres can support a productive farm-to-market business.

The appropriate density depends on regional context; what makes sense in rural Virginia might not be the right density in Montana. In places close to major cities, five units per acre might make sense, while in ranch lands in the West, one unit per 160 acres might be appropriate.

Appropriate density also depends on the community's proximity to cities and to agricultural or natural resource areas. Rural communities on the periphery of cities usually need to accommodate growth, so they need to determine the right density to make sure that the inevitable development is done in a way that enhances the entire community. In communities that are surrounded by open space and that are not experiencing much growth, the edge can be a transition zone where clustered homes on small lots give way to agricultural uses.

A variety of factors fuel low-density development, including:

- People want to move to rural communities for the quality of life.

- Many people want affordable second and vacation homes in rural areas.

- Rural communities want to grow and to generate jobs.

- Greenfield land typically can be developed easily under current zoning with no special approvals.

Dispersed development typically features single-use pods of homes or commercial uses that are not connected to other places. These places lack a town center with a concentration of other uses. To convert these areas into a pattern that can thrive over time, rural communities could designate small town centers. Directing development to those centers could reduce travel between spread-out housing subdivisions or could at least shorten the driving time between locations. These clusters of more intense development with a mix of uses will become gateways to the homes and businesses located nearby.

RESPONSE TO THE PROBLEM

As discussed above, densities that are inconsistent with community character in rural areas create a development pattern that can be worrisome from fiscal, environmental, social, and health perspectives. Developments that provide transportation options, opportunities to access a range of businesses, and access to open space are more likely to sustain themselves over time by attracting and retaining businesses and residents and by using resources efficiently. A community should determine what type of place it is trying to be and then plan for development patterns and associated densities accordingly. There is no specific formula or metric to apply. Addressing this issue is a nuanced process that requires understanding that density ultimately characterizes an area, no matter what a future land use map might indicate. For example, if subdivisions with typical suburban densities are proposed and built, they will likely attract similar densities and commensurate uses, such as commercial shopping strips. Connecting development decisions to the plans that have been developed will help ensure that the community gets the type of development it envisions.

One way to deal with this density context challenge is for communities to make sure that their local comprehensive plans direct new development to areas that are within a natural edge to the community. For example, a major road or a river might provide a barrier to expansion and clearly define an edge to the community.

Another idea for addressing the density context is to expand the town's street pattern (often a terrain-modified grid) while using existing infrastructure capacity, with development ending at an agricultural zone on the community's edge. Some communities reinforce this approach by limiting utility extensions and prohibiting septic systems in the undeveloped land beyond the edge of town. This process will be most effective once the community has committed to this development pattern, as it can be continued outside of the core boundaries of the town and extended to create a consistent density.

These remedies address only the properties at or near a town's edge. Equally challenging are subdivisions and large, freestanding residential and commercial developments scattered in more remote rural areas. These developments are usually under county purview, so dealing effectively with them requires cooperation between municipalities and counties. In these cases, it is important to a town to have a strong relationship with the county government to ensure that there is consensus on how to plan for new development. For instance, questions that will need to be addressed might include: Will the town's development densities be continued in targeted areas in the county to create consistency? What are appropriate densities for transition areas that are acceptable to both the town and county? Answers to these questions require discussion and information exchange.

To get public support to implement changes, communities might need to educate municipal staff and officials, the general public, and other stakeholders about the advantages of more compact development—for example, making stores, schools, parks, and other amenities more economically viable and easier for residents to get to by putting them closer to homes; economies of scale in providing services; and fiscal responsibility. Outreach is typically most effective when it is part of a broader community or regional planning process. Education and understanding can help develop the political will to adopt and enforce zoning codes, development policies, and incentives that will encourage the desired development patterns.

EXPECTED BENEFITS

- Having densities set in advance for designated growth areas gives landowners and developers more predictability.

- More compact development reduces taxpayer costs for local government-provided infrastructure and services.

- Compact development accommodates more growth in developed areas, helping to preserve large contiguous blocks of open space, agricultural lands, and natural resource areas such as wetlands and wildlife habitat.

- Compact development reduces interference with agricultural operations and helps keep farming and ranching viable in the community.

- Development that is compact and well-connected makes walking and biking more appealing, which can make it easier for people to work activity into their daily lives and improve their health.

- Shorter driving distances and more transportation options help reduce greenhouse gases and other pollution.

STEPS TO IMPLEMENTATION

1. Modest Adjustments

- Develop design regulations that require street connectivity with adjacent neighborhoods, and create land use district transitions to adjacent agricultural or undeveloped areas.

- Allow cluster or conservation subdivisions at the edge of town to transition to true rural areas (see Chapter 7: Use Cluster Development to Transition From Town to Countryside).

- Designate locations for small hamlets in rural areas to serve as local service centers for residents. Focus public efforts and programs such as outreach from the chamber of commerce for small business development on these centers to help develop viable small businesses and services.

- Prioritize public works improvements and investment in existing town business districts. Create incentives to encourage well-designed development adjacent to town to make the best use of these investments.

2. Major Modifications

- Adopt town and county comprehensive plans that recommend appropriate densities in town influence areas.

- Establish community service areas in comprehensive plans that limit service provision to towns and town influence areas.

- Adopt true agricultural zone districts (one unit per 20 to 80 or more acres). The size of these districts can vary somewhat depending on geographic region, sites, soils, and the type of agricultural business. Encourage use of conservation easements in these districts.

- Require minimum densities in areas designated for growth.

- Require cluster or conservation subdivisions to be located at the town's edge to provide transition to rural areas. Do not allow them in active agricultural areas or in sensitive natural areas outside town influence areas.

- Revamp the annexation policy to support appropriate densities on the periphery of growth areas. Depending on local context, communities annex land to expand the tax base or to ensure that a particular area is developed in a specific manner once zoning is applied (see Chapter 8: Create Annexation Policies and Development Standards That Preserve Rural Character). Many peripheral areas that could later be annexed are developed with densities that are not appropriate to the character of the area.

3. Wholesale Changes

- Undertake joint town-county planning to develop consistent growth management policies that designate preferred growth areas and limit the use and location of large-scale PUDs and new rural towns in unincorporated areas outside town influence areas.

- Create a review process to ensure that new developments are balanced communities providing a full range of services, housing, and employment, rather than isolated subdivisions.

- Adopt an adequate public facilities ordinance (where permitted by state code) that sets criteria for utility expansion and service of outlying developments, and require areas that fail to meet public facility standards to be prioritized in local capital spending plans. Require that infrastructure, such as roads, water and sewer service, and schools, be in place when new development is constructed.

PRACTICE POINTERS

- Analyze whether existing zoning and subdivision provisions allow division of land for residential development without subdivision review. Piecemeal subdividing without review opens the door for development in rural areas that fragments agricultural or natural lands over time.

- The appropriate lot size in agricultural zone districts will vary depending on the region, state, land use patterns, and types of agriculture. Close to urban markets, smaller lots can be appropriate, generally if agricultural zoning and tax exemption requires proof of active agricultural use.

- Some local governments have provided support for land trusts to purchase or accept donation of conservation easements from farmers and ranchers, allowing landowners to realize some value while maintaining agricultural operations.

- Public outreach and education—using meetings, workshops, and development charrettes—are important to implementing these significant changes.

EXAMPLES AND REFERENCES

Bowers, D. "Achieving Sensible Agricultural Zoning to Protect PDR Investment." Presented at the Protecting Farmland at the Fringe conference, September 6, 2001. http://www.farmlandinfo.org/documents/29520/Achieving_Sensible_Agricultural_Zoning_full_presentation.pdf.

Burchell, R. et al. *Cost of Sprawl –2000*. TCRP Report 74. Transportation Research Board. 2002. pp. 56-80. http://onlinepubs.trb.org/Onlinepubs/tcrp/tcrp_rpt_74-a.pdf.

County of Marin, California. "Marin Countywide Agriculture Element – Executive Summary." http://www.co.marin.ca.us/depts/cd/main/comdev/advance/cwp/ag.cfm. Accessed January 8, 2010.

Daniels, T. "What to Do about Rural Sprawl?" Presented at the American Planning Association Conference, Seattle, WA. April 28, 1999. http://www.mrsc.org/subjects/planning/rural/daniels.aspx.

Duerksen, C. and Van Hemert, J. *True West: Authentic Development Patterns for Small Towns and Rural Areas*. American Planning Association. 2003.

Freedgood, J., Tanner, L., Mailler, C., et al. *Cost of Community Services Studies: Making the Case for Conservation*. American Farmland Trust. 2004. http://www.farmlandinfo.org/documents/27757/FS_COCS_8-04.pdf.

Freedgood, J. *Saving American Farmland: What Works*. American Farmland Trust. 1997. http://www.farmlandinfo.org/farmland_preservation_literature/index.cfm?function=article_view&articleID=29384.

Livingston, A., Ridlington, E., Baker, M. *The Costs of Sprawl: Fiscal, Environmental, and Quality of Life Impacts of Low-Density Development in the Denver Region*. Environment Colorado. 2003. http://www.policyarchive.org/handle/10207/5153.

Pruetz, R. *Beyond Takings and Givings*. Arje Press. 2003.

U.S. Department of Agriculture. Farmland Protection Policy Act. http://www.nrcs.usda.gov/wps/portal/nrcs/detail/national/programs/alphabetical/fppa/?&cid=nrcs143_008275. Accessed December 11, 2009.

Washington Department of Community, Trade and Economic Development. *Keeping the Rural Vision: Protecting Rural Character & Planning for Rural Development*. 1999. http://www.cted.wa.gov/DesktopModules/CTEDPublications/CTEDPublicationsView.aspx?tabID=0&alias=CTED&lang=en&ItemID=974&MId=944&wversion=Staging.

Wells, B. *Smart Growth at the Frontier: Strategies and Resources for Rural Communities*. Northeast-Midwest Institute. 2002. http://www.activelivingbydesign.org/events-resources/resources/smart-growth-frontier-strategies-and-resources-rural-communities.

7 USE CLUSTER DEVELOPMENT TO TRANSITION FROM TOWN TO COUNTRYSIDE

INTRODUCTION

Cluster or conservation development[26]—homes clustered on a portion of a site and the rest of the land preserved as open space—is used to preserve large tracts of open space and agricultural land. Clustering allows landowners and developers to attain the overall allowable density on a site—getting the most development potential out of the site—while preserving a significant amount of it as open space. While clustering can be an effective tool, many rural jurisdictions do not get the results they expect.

If they are near agricultural lands, cluster developments can introduce residents into the area who might not be used to living near farming operations. Complaints about noise, dust, and odors; harassment of livestock by domestic pets; and other issues often follow. Nearby farms might be forced to take expensive mitigation measures or even shut down. Similarly, cluster developments in ecologically sensitive areas can fragment wildlife habitat, introduce invasive species to the detriment of others, and introduce humans and pets into the habitat. For these reasons, cluster developments should be carefully located.

Cluster developments work best where towns transition to true rural areas with active agricultural or livestock operations and large contiguous natural areas. In transition areas, the developed cluster can be adjacent to existing development on the edge of town, with the open space acting as a transition or buffer that separates the development from undeveloped areas. This approach can work as long as extensive additional growth is not expected; otherwise, that additional growth could leapfrog to the other side of the cluster buffer with limited connections to the town.

Photo courtesy of Victoria Ranney

Cluster development can help a rural community transition between town and countryside. Prairie Crossing in Grayslake, Illinois, clustered homes to protect a large swath of prairie. The community includes a station on a rail line that goes to Chicago, a working farm, historic community buildings, and energy-efficient new homes.

---

26 These terms are nearly interchangeable. For the purpose of this chapter, only *cluster developments* will be used.

Cluster developments are often stand-alone subdivisions in the countryside surrounded by open space, unconnected to towns and requiring residents to drive long distances to get to daily destinations. Learning from this experience, local governments are beginning to direct cluster development to the periphery of existing towns and villages or are limiting their size (e.g., no more than 10 residential lots) to control the impact they have on rural character, agricultural operations, and wildlife habitat. However, even with these strategies, cluster developments can create concentrations of homes in locations so spread out that residents still must drive everywhere.

RESPONSE TO THE PROBLEM

As a first step, small towns and rural counties can adopt zoning and subdivision provisions that allow cluster development only at the periphery of towns. Rural local governments often resist smaller lots (e.g., less than 2 acres) in rural areas, assuming that they will erode rural character. However, when cluster developments are used in appropriate locations—areas between towns and true rural areas—they can provide a smooth transition between town-scaled development and open lands. The homes can be adjacent to already-developed areas (to provide connectivity) or areas with an available mix of uses, infrastructure, and services, while the open space portion of the site provides a buffer between the built-up area and rural land.

To use cluster development effectively, communities need to decide which transition areas are most appropriate for this approach. Offering zoning and/or development incentives can encourage development in those locations. By mapping areas that should be preserved as working lands or natural resource areas and areas that could support future infrastructure expansion, the community can direct development to locations that make sense. Requiring open space preserved through cluster development to abut existing open spaces protects large blocks of land, which better supports agriculture, wildlife habitat, and rural landscapes over the long term.

Some communities mandate standards for cluster development in their ordinances. Others offer voluntary cluster development ordinances with incentives, such as density bonuses. Density bonuses can be flexible, with the number of additional units based on the quality of the design or other community benefits. Clustering offers the most benefits to the community when

Serenbe, a development about 30 miles southwest of Atlanta, Georgia, preserves more than 70 percent of its land as farmland and natural green space. It clusters development into three hamlets that include various housing types, restaurants, live-work spaces, stores, and services.

Photo courtesy of UGArdener via Flickr.com

cluster development locations are chosen based on local and regional priorities for preserving natural habitat and cultural treasures. Communities could measure how well a proposed cluster development meets specific, measurable factors such as:

- The per unit amount of impervious surfaces, road length, or loss of woodlands and other specific resources.

- Orientation of lots around a central green or square or an amenity such as a meadow, a stand of trees, a lake, or another natural feature.

- Preservation of visually prominent areas such as ridges or hilltops and areas along secondary public roads.

- Reducing peak discharges of stormwater runoff to levels that consistent with the discharges from that site before it was developed.

- Capture of 80 percent of the sediments and pollutants in runoff from a one-year storm event.

EXPECTED BENEFITS

- Well-designed and -located cluster development can provide an appropriate transition between town and countryside.

- Cluster development can permit ranchers, farmers, and other landowners to realize development value from their property while protecting large, contiguous blocks of open space for agriculture or to protect sensitive natural areas.

- Local governments can avoid fragmentation of agricultural lands and wildlife habitat when they approve cluster development in preferred locations inside town influence areas.

- Compact, well-designed cluster development requires less paved area for roads and less expansion of water and sewer infrastructure.

- Cluster development can provide environmental and fiscal advantages, such as reducing infrastructure costs and making it cheaper to provide community services (e.g., police and fire protection).

STEPS TO IMPLEMENTATION

1. Modest Adjustments

- Require open space, agricultural, and/or ranchland preservation plans on the development site as part of a cluster development proposal.

- Create a comprehensive cluster development policy, summarizing the community's vision for land uses, connectivity to the existing town, and natural resource preservation for new development proposals.

- Provide modest density bonuses to encourage cluster development in town influence areas (e.g., one additional unit for every 10 units permitted under current zoning).

- Allow community septic systems for cluster developments in town influence areas where central sewer is not available.

2. Major Modifications

- In comprehensive plans, designate growth areas that are appropriate locations for cluster development.

- Adopt comprehensive cluster development regulations as an alternative to standard development in all zone districts on the town's edges.

- Adopt future development standards so that clusters in town influence areas can accommodate more development and get infrastructure in the future (e.g., provide easements for water and sewer lines and drainage or designate future connections for rights-of-way to create a connected street network).

3. Wholesale Changes

- Require open space, agricultural, and/or ranchland maintenance and management plans for all cluster development.

- Prohibit cluster development in viable agricultural and sensitive natural areas. Designate prohibited locations in the land use plan and on the zoning map.

- Mandate that permit approvers use specific performance criteria in reviewing and approving cluster subdivision proposals.

PRACTICE POINTERS

- In drafting cluster subdivision provisions, specify preferred locations for open space (e.g., environmentally sensitive areas). Encourage sites that are contiguous with existing development, but allow non-contiguous open space in specific, defined circumstances (e.g., where there are multiple natural features on a site such as streams and steep slopes).

- During the planning phases, ensure the development includes open space, preserves views, and limits impacts on natural areas as required by the local jurisdiction.

- Reach out to landowners and developers to educate them about the process and the benefits of cluster development, especially the potential tax advantages of putting easements in place.

EXAMPLES AND REFERENCES

Arendt, R. *Conservation Design for Subdivisions: A Practical Guide to Creating Open Space Networks.* Island Press: Washington, DC. 1996. pp. 33-38.

Church, J. "Local Community Resources: Cluster/Conservation Development." University of Illinois Extension. http://urbanext. illinois.edu/lcr/LGIEN2000-0010.html. Accessed January 8, 2010.

Duerksen, C. and Snyder, C. *Nature-Friendly Communities: Habitat Protection and Land Use Planning.* Island Press: Washington, DC. 2005. "Chapter 4: Baltimore County, MD: Using the Whole Toolkit for Habitat Preservation."

Haines, A. "Regulatory Approaches to Conservation Subdivisions in Wisconsin." *The Land Use Tracker*, University of Wisconsin-Stevens Point, Center for Land Use Education, vol.2, no.1. 2002. http://www.uwsp.edu/cnr/landcenter/tracker/ Summer2002/Tracker.html.

Ipswich River Watershed Association (Massachusetts). *Water Wise Communities: A Handbook for Municipal Managers in the Ipswich River Watershed.* 2006. http://ipswich-river.org/ resources/water-wise-communities-handbook.

Ohm, B. *An Ordinance for a Conservation Subdivision.* University of Wisconsin Extension. 2000. http://urpl.wisc.edu/ people/ohm/consub.pdf.

 CREATE ANNEXATION POLICIES AND DEVELOPMENT STANDARDS THAT PRESERVE RURAL CHARACTER

INTRODUCTION

Communities often have the most control or influence over development on their edges when they annex those areas. Communities can determine how annexed land can help advance the community vision and planning goals and ensure that public costs of developing annexed areas (including infrastructure capital and operating costs and public services) are balanced with potential tax and other revenues.

Because many rural communities have resource constraints, they might not have the capacity to effectively evaluate all proposed annexations. Few have adopted annexation policies that are coordinated with their comprehensive plans and growth strategies. Nor have most rural towns reached agreements with surrounding or adjacent counties and townships regarding town-level residential and commercial development proposed in surrounding unincorporated areas. Such agreements typically require the proposed development to explore annexation with the adjacent town or village prior to receiving approvals or to agree not to object to future annexation requests by the town. Without evaluation standards, annexation policies, or interjurisdictional agreements, the result is often spread-out or scattershot rural developments that drain local government coffers, strain government service and infrastructure providers, and contradict local comprehensive plans and community goals.

Over time, rural small towns often become financially overwhelmed by providing services to low-density, spread-out developments in surrounding unincorporated areas. This pattern typically occurs when development is allowed on large lots— one unit per 2 or more acres—that use wells and septic systems rather than centralized water treatment. Local governments might find they cannot annex and develop these areas because there are no easements to run water and sewer lines; rights-of-way and street linkages are inadequate to build a grid of town streets; and the scattered, large-lot pattern makes village-scaled developments nearly impossible. As a result, pressure mounts

Vienna, Maryland, annexed a large parcel of land (outlined in red) in its designated growth area. Two-thirds of the parcel is protected open space that creates a greenbelt and provides buffers for waterways and for farmland. The remaining land can be developed but must connect to the town; one potential concept for this development is illustrated in this plan. Building and street design guidelines, architectural standards, and other guidelines will help the new development fit with Vienna's character. The goal is for the new neighborhood to become a true extension of the town.

for development that can leapfrog the low-density, spread-out developments.

One of the most important forces driving annexation is the desire of cities and towns to increase their tax base and revenues. In areas with multiple jurisdictions that are experiencing growth, municipalities also find that if they do not annex aggressively, they might be hemmed in by others' annexations, thus limiting their ability to expand. Municipalities might also believe the only way to ensure that growth in the surrounding region occurs responsibly is to annex areas to gain control over planning, development, and design decision-making before development occurs.

However, jurisdictions need to be thoughtful about the long-term implications of annexation. In some cases, public expenditures on annexed areas can exceed increased tax revenues from these areas, especially over the long term. This imbalance is often true of lower-density development added near—but not contiguous to—existing communities, which requires road improvements and infrastructure extensions. Even if a development pays the full initial costs of infrastructure improvements—and many states do not allow communities to require such payments—the increased operating, maintenance, and service costs of more dispersed development still can have a major long-term impact on the community's budget (see Chapter 2: Incorporate Fiscal Impact Analysis in Development Reviews).

RESPONSE TO THE PROBLEM

Rural communities can consider the following policies to improve the annexation process and ensure that annexed areas meet the community's development standards:

- Revise local codes to require that annexations be included in the comprehensive planning process.

- Develop intergovernmental processes and agreements— building partnerships between counties and municipalities and between neighboring municipalities—to guide and govern planning and funding for expansion and annexation.

- Establish criteria and a standard review process for potential annexations, including criteria for fiscal impact analyses; required road and infrastructure connections; assessing the need for parks, open space, schools, and other community facilities; and development standards.

- Develop an integrated approach to make sure that annexation is concurrent with adopted zoning and development standards for required infrastructure and community facilities.

- Provide early and frequent opportunities for meaningful citizen participation in annexation and development decisions.

In addition to consideration of development-specific fiscal impacts, annexation review should involve assessment of

the community's overall infrastructure capacity—regional transportation, water supply, sewers, schools, parks, fire stations, and other civic facilities. This underlying needs and capacity analysis can help determine what kinds of facilities will be required in areas to be annexed and can be a starting point for negotiations, proffers, or exactions from individual developments (depending on state laws).

Because ad hoc annexation is often driven by local competition for tax revenue, communities could also choose to work with nearby jurisdictions to coordinate their local taxation systems. Revenue sharing among jurisdictions, where allowed by state statute, is one potential solution. Intergovernmental cooperation could also include working together as a coalition to apply for federal and state economic and community development funds. In some states, towns and counties sign intergovernmental agreements to apply town standards in town influence areas. In others, state law gives municipalities the authority to impose their subdivision standards on county subdivisions around their borders. Some local governments draft joint land use plans between towns and counties for areas around towns and adopt joint land use regulations to ensure that new development meets town standards.

Successful use of annexation requires the coordination of partnerships among neighboring local governments, residents, environmental groups, businesses, and developers. These partnerships are frequently an outgrowth of a regional planning process that creates a shared vision of how and where the community should grow and what it should look like in the future (see Chapter 1: Determine Areas for Growth and for Preservation). A shared vision can help rural towns reach agreements with surrounding and adjacent counties to require that town zoning, subdivision standards, and design guidelines be applied to new developments in designated growth areas outside the town's borders. This collaboration could result in development with a better-connected network of roads, wider rights-of-way, and reserved or dedicated connection points to accommodate more compact future development when that development is annexed into the adjacent town. In some areas, towns and counties have reached agreements that require developments in unincorporated areas to include language in deeds or homeowners' association agreements stating that residents agree not to object if the town wants to annex the development in the future.

One strategy to ensure that areas to be annexed are compatible with the existing community is to create a plan for annexation based on the patterns and character of adjacent neighborhoods. To define the desired development type more specifically, communities can adopt a unified development ordinance that brings together subdivision and zoning ordinances and neighborhood development regulations, including street design guidelines and connectivity requirements, development standards that allow a mix of uses and a variety of home and lot sizes, utility and open space guidelines, and protection of sensitive habitat and cultural resources.

EXPECTED BENEFITS

- Local governments can secure community benefits, such as open space and infrastructure contributions, during the annexation process.

- Fiscal impact analyses required as part of a community annexation policy will give local governments a more accurate picture of the true costs and benefits of a proposed development in terms of potential tax revenues and costs of services and facilities.

- Annexation agreements avoid intergovernmental competition for territorial expansion that can lead to over-extension of town boundaries and a scattered, leapfrog development pattern.

- Orderly annexation helps preserve rural resources, such as agriculture, open space, stormwater infiltration, working lands, and natural habitat, and maintain a distinction between "town" and "country."

- Annexation policies help avoid the ad hoc formation of small, incorporated municipalities that can hinder the expansion of existing towns. [27]

- Orderly, planned community expansion accommodates population growth and provides the tax base required to meet the community's objectives.

- Subdivisions and commercial development in town influence areas will be built to standards that make it easier for the properties to accommodate new development or to be annexed into the town in the future.

- Uniform town-county standards in town influence areas help to create predictability regarding community expectations.

- Uniform standards based on joint planning will help produce rational settlement patterns that preserve the ability of the town to expand in a logical fashion, thereby helping to prevent inefficient leapfrog development.

- Better planned, more functional town centers can emerge, serving larger areas more efficiently. In addition, the area can attract a greater, more diverse mix of amenities, stores, services, and job opportunities.

STEPS TO IMPLEMENTATION

1. Modest Adjustments

- Encourage future annexations to be consistent with the community comprehensive plan (or local equivalent) and require that the comprehensive plan maps and describes future potential areas of annexation.

- Encourage future potential annexation areas mapped in the comprehensive plan to include a preliminary identification of anticipated zoning as well as a preliminary analysis of how municipal services and infrastructure (e.g., water, sanitary sewer, stormwater, transportation, and police and fire) would be funded. This analysis should be based on community service standards and an assessment of existing conditions and revenue capacities in the mapped areas.

- Encourage mapping of potential future annexation areas in the comprehensive plan to identify and evaluate prime agricultural lands, important wildlife habitat, areas of special ecological value or concern, and lands contaminated by past agricultural or industrial activities.

- Establish a code requirement that the transportation element of the community comprehensive plan (or local equivalent) identify a future network of streets connected with the existing town pattern for any potential future annexation areas mapped in the plan. Require that extensions of the existing street network be mapped to meet minimum internal connectivity standards within any annexed areas, as well as external connections with existing and future neighborhoods and developed areas.

- Require annexation proposals to be accompanied by a site plan with enough specificity to allow the local government to undertake a fiscal impact analysis.

---

27   Towns sometimes incorporate to avoid being subject to taxes imposed by a neighboring jurisdiction to pay for municipal services.

Photo courtesy of USDA Natural Resources Conservation Service

In Sonoma County, California, the Local Agency Formation Commission (LAFCO) reviews and approves proposed annexations. LAFCOs were created by state law to coordinate local government agencies and protect farmland.

- Encourage communities to work together as a coalition to potentially gain an advantage in seeking federal and state economic and community development funding.

- Encourage towns and counties to undertake joint land use planning in town influence areas, to adopt plans designating growth areas, and to establish similar development quality and improvement policies.

- Encourage counties to require new development in town influence areas to meet the town's subdivision ordinance and other development standards (e.g., street design guidelines and connectivity requirements, development standards, utility guidelines, and design guidelines) or to be capable of upgrading to meet such standards upon annexation.

2. Major Modifications

- Adopt detailed fiscal impact analysis requirements for proposed annexations, including criteria for comparing revenues to costs. Include provisions for additional fees and funding to rectify imbalances where costs outweigh revenues. Include provisions for special cases where annexation of lands can be justified based on other community objectives (e.g., protecting open space, recreational lands, or water supplies).

- Establish a minimum contiguity requirement for any proposed annexation area depending on the physical character of the site. A sample requirement might be that at least 25 percent of the circumference of any proposed annexation must be coterminous with the existing incorporated area, subject to exceptions for bodies of water, public parks, or other similar features. An adjunct provision or variation would be to prohibit "flagpole"[28] annexations.

- Develop and adopt joint infrastructure standards (for water, sanitary sewer, stormwater, and streets) for use by a municipality and a surrounding or adjacent county or by multiple municipalities and/or counties to be applied to proposed development in areas that the parties have agreed could eventually be annexed into a municipality. These standards ensure that development in future annexation areas is designed to be consistent with the municipalities' standards.

---

28   A flagpole annexation is a parcel that is connected to a larger entity, such as a municipality, by a narrow strip of land.

- Require that annexed parcels be zoned in accordance with the adopted comprehensive plan.

- Develop an intergovernmental agreement between one or more municipalities and one or more counties to guide the annexation process in potential annexation or growth areas mapped in the agreement. Include provisions addressing infrastructure standards, funding of infrastructure and services, and approval processes of the affected jurisdictions.

- Build on any joint town-county plans for town influence areas, and adopt uniform zoning and subdivision standards by intergovernmental agreement.

3. Wholesale Changes

- Where allowed by state law, the town and county could form a joint planning commission to undertake development reviews and apply uniform standards in town influence areas.

- Develop an intergovernmental agreement between one or more municipalities and one or more counties providing for development and adoption of a regional and multijurisdictional comprehensive plan. Include provisions for identifying areas of potential future annexation and provisions for zoning, infrastructure, lands of special concern, and street extensions.

- Develop a regional compact or intergovernmental agreement for revenue sharing to reduce or eliminate the pressures to annex land for municipal budget growth. Include a "fix it first" component in the agreement to ensure that existing facilities and infrastructure are not abandoned or allowed to further deteriorate in favor of new development in annexed areas.

PRACTICE POINTERS

- Annexation law and policy are among the most controversial aspects of growth management. Several states are currently legislating on the subject of annexation—changing laws governing municipalities' authority to annex land, establishing or revising criteria for annexations, requiring additional review and approval by adjacent counties and municipalities, or providing for oversight by third parties or agencies. The first step for any municipality

is to make sure that existing and proposed local ordinances are consistent with state law.

- Issues related to estimating the costs of extending infrastructure and services into potential annexation areas are difficult to resolve if there are no agreed-upon standards for the timing, placement, and design of facilities and services. Establishing the design and service standards that will be used to estimate the cost of providing facilities and services—ideally in cooperation with other area governments—will help localities make rational and consistent annexation decisions.

- One potential benefit of good annexation policy, especially with multiple jurisdictions involved, is avoiding the leapfrogging of residential and commercial development into rural areas. This benefit will not be realized if the county continues to permit development that is not rural in character. Changes to county zoning and land development codes are an essential component in a rational annexation process.

- To support small towns and rural counties, which typically have limited planning and development staff, state and regional organizations can compile a list of federal funding resources that can be used as incentives, or "carrots," to counter what might be perceived as the "stick" of limitations under revised annexation policies.

- Joint planning efforts typically require significant public involvement and education to ensure that residents of both the town and county, especially those in the town influence area, have a chance to influence decisions. These efforts are important in areas facing growth pressures as well as in older areas with little growth, where the town is declining and the limited growth in that area is moving into surrounding greenfields.

EXAMPLES AND REFERENCES

Boulder County, Colorado. *Boulder County Comprehensive Plan*. http://www.bouldercounty.org/government/pages/bccp.aspx. Accessed February 22, 2012.

Colorado Office of Smart Growth. *Planning for Growth: Intergovernmental Agreements in Colorado*. September 2006. http://cospl.coalliance.org/fez/eserv/co:3186/loc61202p692006internet.pdf.

Denver Regional Council of Governments. *Mile High Compact.*
http://www.drcog.org/index.cfm?page=MileHighCompact.
Accessed January 7, 2010.

Edwards, M. "Understanding the Complexities of Annexation."
Journal of Planning Literature. Vol. 23, No. 2, 119-135. 2008.
http://jpl.sagepub.com/cgi/content/abstract/23/2/119.

Hinze, S. and Baker, K. *Minnesota's Fiscal Disparities
Programs.* 2005. http://www.house.leg.state.mn.us/hrd/pubs/
fiscaldis.pdf.

Local Agency Formation Commission of Monterey County,
California. *Policies and Procedures Relating to Spheres of
Influence and Changes of Organization and Reorganization.*
Adopted April 25, 2011. http://www.co.monterey.ca.us/
lafco/2011/WEB%20POSTS/OLD/June%2016/Policies%20
and%20Procedures%20April%2025%202011.pdf.

Larimer County, Colorado. *Larimer County Urban Area Street
Standards.* Revised April 2007. http://www.co.larimer.co.us/
engineering/GMARdStds/GMARdStds.htm.

Larimer County, Colorado. "Rural Land Use Center." http://
www.co.larimer.co.us/rluc/. Accessed January 8, 2010.

Nelson, A. *Urban Containment in the United States.* American
Planning Association. April 2004.

Town of Berthoud, Colorado. "Town of Berthoud/Larimer
County Intergovernmental Agreement." Executed August 22,
2000. http://www.co.larimer.co.us/planning/planning/berthoud_
iga.pdf.

Town of Vienna, Maryland. *2003 Town of Vienna Comprehensive
Plan—2009 Comprehensive Plan Amendments.* September 2009.
http://www.viennamd.org/2009_gvcomp_revision.pdf.

Trohimovich, T. "How the Growth Management Act Changed
Annexation & Current Issues in Annexation." 1000 Friends
of Washington. 2004. http://www.futurewise.org/resources/
publications/Annexation.pdf.

PROTECT AGRICULTURAL AND SENSITIVE NATURAL AREAS

INTRODUCTION

Sensitive natural areas such as wetlands, wildlife habitat, beaches, and steep slopes are important from an environmental perspective, but they also help create the special character of rural areas. They are often important contributors to the local economy, bringing tourism, providing ecosystem services like protecting water quality, and supporting the health of working farmland, forests, and fisheries.

Rural local governments know that working lands, farms, prairies, forests, and rangelands are central to both their heritage and their economic future. Working lands are often at the heart of communities' distinctive rural character—and are often the reason the towns were settled in the first place. Many rural places depend economically on traditional resource industries, such as agriculture, forestry, and mining, and related processing, manufacturing, and trade. In a successful rural economy, a healthy balance can be maintained between the tourist and resource sectors, such as a vineyard that includes a restaurant and a shop, or an orchard with a cider mill and a catalog store operation. Developing supportive policies, land use regulations, and zoning that allow an "agricultural workplace" category can help keep families on the farm and prospering.

RESPONSE TO THE PROBLEM

Jurisdictions are adopting a variety of protective regulations, land use planning policies, land acquisition programs, density transfer programs, and land preservation programs to protect sensitive natural areas and wildlife habitat, as well as to preserve and maintain farmland. The actual or speculative loss in value that occurs when a local government enacts land use regulations to protect land can cause controversy and could spawn legal action. In response, local governments have turned to tools and techniques that provide options for landowners to recoup some of the land value that might be diminished, or perceived to be diminished, by regulations.

Under the zoning code in Kailua-Kona, Hawaii, farms can offer tours of their facilities and sell coffee to the public.

Two relevant programs are purchase of development rights (PDR) and transfer of development rights (TDR). PDR and TDR programs can help gain new support for land protection strategies in rural areas by offering some compensation to affected landowners to offset their potential loss in value. In concept, PDR and TDR programs are simple. A typical rural property identified for possible preservation, which contains high-value natural resource areas or agricultural lands, could be zoned for 1- or 2-acre-lot residential development. To protect the land under a PDR program, the local government would appraise the value of the development rights on a parcel and then purchase a conservation easement that either prohibits development or allows it only at a lower density. Public access to the preserved land might or might not be part of the transaction. Funding for the PDR program might come from general tax revenues, an open space bond issue, or a dedicated funding source such as an earmarked sales tax. The owner typically stays on the property and continues to use the land as he or she did prior to the agreement.

Under a TDR program, the local government classifies property as sensitive land or agriculture through tools such as agricultural zoning or sensitive lands protection regulations, putting much of the land off-limits to development. This action turns such properties into "sending areas." To reduce the financial impact on the sending-area landowner, the local government allows the landowner to sell his or her development rights to a developer who wants to build in a designated growth area—the "receiving area." The developer pays the sending-area landowner for those development rights and then has the right to build more than originally designated. If the TDR program is designed correctly, with a clear understanding of how large the sending and receiving areas should be to create a viable market for development rights, it can be an effective tool to protect large tracts of open space and working farmland. Local government staff must pay attention to the mechanics of the process (e.g., how to determine how many development rights are assigned to a particular property and the documentation of the transfer). Successful TDR programs like those in the New Jersey Pinelands[29] and Montgomery County, Maryland,[30] can be an effective melding of regulations and incentives. In many jurisdictions, this combination could be more appealing than regulations alone.

Other financial tools that help make it possible for landowners to keep farmland in production and avoid the need to sell land include federal, state, and local conservation tax credits, which provide incentives for donating land or conservation easements, and local tax policies, such as use value taxation, which assesses farmland or conservation land at a lower value than it would be worth if sold for development.

Updated zoning can also support job creation that considers social and environmental impacts while preserving working farms and lands, especially smaller farms that can become surrounded by development. Older zoning might not allow commercial, light manufacturing, retail, or related uses in an agricultural zone. A new "agricultural workplace" zone could allow those uses on an owner-occupied farm, allowing economic development activities, home offices, on-farm sales, and agriculture-related industry.

29   New Jersey Highlands Water Protection and Planning Council. "Established TDR Programs in New Jersey." State of New Jersey Department of Agriculture. 2007. http://www.nj.gov/agriculture/sadc/tdr/casestudy/tdrexamplesnj.pdf.

30   Montgomery County, Maryland, Department of Economic Development. "TDR Program Overview." 2006. http://www.montgomerycountymd.gov/content/ded/agservices/pdffiles/tdr_info.pdf.

EXPECTED BENEFITS

- Preserving natural resources contributes to local economies by bringing tourism, hunting, fishing, and other recreational uses.

- Protecting working lands and farms contributes to the economy and rural character while preserving property values.

- Preserved areas tend to cost local governments less than they produce in taxes, due to lower demand for costly town-level services when land remains undeveloped.

- TDR programs that direct development to designated growth (receiving) areas preserve open space, reduce fragmentation of sensitive natural areas, and reduce opposition to agricultural and sensitive lands protection programs.

- TDR receiving areas allow more cost-effective delivery of government-funded infrastructure and services and focus development to attract more consumers, services, and commercial development.

- Preserving agricultural lands and jobs supports agriculture-related economic development that is sustainable over the long term.

STEPS TO IMPLEMENTATION

1. Modest Adjustments

- Identify and map sensitive natural resources.

- Adopt policies to protect these resources, including limiting capital improvements (such as road improvements or extending water and sewer lines beyond certain developed areas) that might lead to development or degradation. Include opportunities to preserve individual sensitive natural areas in rural towns that connect to larger environmentally sensitive areas and open space in the countryside.

- Seek assistance from state natural resource agencies in development reviews and assessment of impacts on sensitive natural areas. Require larger projects to provide funding that will allow the local government to retain a consulting planner or resource biologist, or charge sufficient application fees to pay for such reviews.

- Establish government service boundaries to encourage in-town development. Demonstrate the cost of service provision outside these boundaries to property owners.

- Work with local land trusts to help secure conservation easements, provide technical assistance, and explain to potential donors the process and the benefits they might realize from pursuing a conservation easement.

- Enact protective regulations such as development setbacks from rivers and a development setback from streams, wetlands, and lakes.

- Seek economic and community development grants. These grants can allow local officials to offer financing incentives and technical assistance to channel commercial and industrial growth to in-town, infill locations and away from sensitive habitat areas, conserving open space while encouraging economic and job growth.

- Fund a PDR program annually out of general fund or other designated revenues. Work with water and drainage districts to use utility and other available fees or taxes for targeted acquisitions (e.g., buying riparian habitat around a lake to protect water quality). Purchase land identified as sensitive natural areas in the comprehensive plan.

- Institute property tax relief or freeze for properties that maintain rural character in the face of development pressure to make sure that surrounding development does not increase land valuation to a point where property owners feel compelled to sell.

- Incorporate tax increment financing (TIF)[31] districts in receiving areas to help fund both the new, compact development in the receiving areas and the activities and services needed in the preserved natural areas.

2. Major Modifications

- Hire staff or part-time consultants with a resource biology background to help assess plans and development proposals.

- Adopt zoning district requirements (e.g., lot sizes) that do not allow significant residential development in sensitive natural areas identified in comprehensive plans.

- Adopt a PDR program with a dedicated funding source (e.g., a large bond issue or an earmarked sales).

- Enact a TDR program. Downzone (reduce permissible density) in sending areas and grant development credits to landowners. Allow new development only in receiving areas through the purchase of development credits.

- Adopt agricultural workplace zoning districts.

- Purchase natural resource areas such as wildlife habitat and wetlands (or purchase development rights) to protect them from future development.

- Adopt a TDR or PDR program to protect designated sensitive natural areas and transfer density to designated growth areas. Make sure the TDR or PDR initiative includes information on tax advantages and other incentives linked to conservation easements and similar strategies.

- Purchase key sites and hold them in a land bank[32] for future development. Develop partnerships with community development corporations, housing authorities (especially those with bonding power), nonprofit development companies, and others to raise funds needed to acquire desired sites.

3. Wholesale Changes

- Develop a resource protection master plan and adopt it as part of the comprehensive plan. Map areas to protect, or conduct surveys to determine boundaries for protection areas.

- Create a permanent source of funding for sensitive area and open space acquisition, such as a sales tax earmark or bond issue. A specific revenue stream, such as a sales tax earmark or user fees, is required to fund a bond option. Another option would be a linked user fee—for example, greens fees from a nearby public golf course—dedicated to funding sensitive area preservation and restoration.

- For places with a PDR program, expand it by fee purchase of sensitive lands and resell the land with conservation restrictions. Such programs tend to need more upfront capital funding and have longer carrying periods but might be more effective in the end because the preserved land can

---

31   Under tax increment financing, communities can capture the additional property tax revenue generated by the higher property values resulting from investment in a designated area. The new revenue is typically used for infrastructure improvements in the designated area or to retire debt. Most, but not all, states use tax increment financing, and each state has its own requirements and laws.

32   Typically, land banking is used to hold land until a time when the market conditions or other community considerations are favorable for that land to be developed. Land banking can also be used to temporarily hold land out of development until it is feasible to combine it with adjacent parcels for a larger development.

be resold to recoup most of the sales price and will still be protected.

- Adopt a regional TDR program with transfers between rural county sensitive (sending) areas and town development (receiving) areas.

- Explore other development rights for TDR purchases in addition to granting more density in receiving areas, such as allowing developers to buy credits to build larger homes or expand water supply infrastructure.[33]

- Require funding for restoration of degraded habitat on development sites. Use open space funds to restore degraded habitat on protected lands (e.g., stream banks damaged by cattle).

PRACTICE POINTERS

- Work closely with the agricultural community to establish habitat protection programs. Where possible, use incentives such as TDR programs and habitat restoration cost-sharing grants.

- Tie PDR and TDR programs to local comprehensive and open space plans that identify high-value agricultural lands and sensitive areas.

- Balance credits available from TDR sending areas with the absorption capability of the receiving areas. Several communities have struggled when the sending areas are too large and too many development credit sellers are chasing too few buyers, which reduces the value of development credits.

- Make sure TDR receiving areas are designed to receive increased development, which should match the locally preferred intensity and height.

EXAMPLES AND REFERENCES

1000 Friends of Florida. *Wildlife-Friendly Toolbox*. http://www. floridahabitat.org/wildlife-manual/wildlife-friendly-toolbox. Accessed January 8, 2010.

Arendt, R. *Conservation Design for Subdivisions: A Practical Guide to Creating Open Space Networks*. Island Press: Washington, DC. 1996. pp. 33-38.

Barnes, T. and Adams, L. "A Guide to Urban Habitat Conservation Planning." University of Kentucky Cooperative Extension Service. 1999. http://www.ca.uky.edu/agc/pubs/for/for74/for74.pdf.

Duerksen, C. and Snyder, C. *Nature-Friendly Communities: Habitat Protection and Land Use Planning*. Island Press: Washington, DC. 2005. "Chapter 4: Baltimore County, MD: Using the Whole Toolkit for Habitat Preservation."

Duerksen, et al. *Habitat Protection Planning: Where The Wild Things Are*. Planning Advisory Service Report 470/471.

Environmental Law Institute. *Conservation Thresholds for Land-Use Planners*. 2003. http://www.elistore.org/reports_detail.asp?ID=10839&topic=Conservation.

Miller, G. and Krieger, D. "Purchase of Development Rights: Preserving Farmland and Open Space." *Planning Commissioners Journal* 53, Winter 2004. http://www.plannersweb.com/wfiles/w140.html.

National Association of Realtors. *Field Guide to Transfer of Development Rights (TDRs)*. http://www.realtor.org/library/library/fg804. Accessed January 8, 2010.

Nolon, J. *Open Ground: Effective Local Strategies for Protecting Natural Resources*. Island Press 2003.

Pruetz, R. *Beyond Takings and Givings*. Arje Press. 2003.

Skoloda, J. "Wildlife and Habitat in a Comprehensive Plan." *The Land Use Tracker*. University of Wisconsin-Stevens Point, Center for Land Use Education. Fall 2002. http://www.uwsp.edu/CNR/landcenter/tracker/fall2002/wildlife.html.

Western Governors' Association, Trust for Public Land, and National Cattlemen's Beef Association. *Purchase of Development Rights: Conserving Lands, Preserving Western Livelihoods*. 2002. http://www.westgov.org/wga/publicat/pdr_report.pdf.

Wright, J. and Skaggs, R. *Purchase of Development Rights and Conservation Easements: Frequently Asked Questions*, Technical Report 34, College of Agriculture and Home Economics, New Mexico State University. http://aces.nmsu.edu/pubs/research/economics/TR34.pdf.

---

33 Pitkin County, Colorado, for example, allows house sizes of more than 5,750 square feet only if the homeowner purchases development credits from sending-area landowners.

PLAN AND ENCOURAGE RURAL COMMERCIAL DEVELOPMENT

INTRODUCTION

Like all economically sustainable places, rural communities need a strong commercial base. A commercial zoning designation typically allows offices, stores, services, restaurants, medical facilities, and similar activities, but not residences. Newer zoning codes—based on patterns long established in nearly every town in America—incorporate a variety of commercial and residential types and uses into mixed-use zoning. A mix of uses reduces driving distances and makes it easier for people to walk or bike to their daily destinations because homes, workplaces, stores, schools, and services are closer together. Directing commercial development to existing towns and villages helps encourage residential growth in town and reduces the likelihood of scattered businesses in rural areas that encourage more spread-out development and fragmented land. Encouraging commercial development in towns helps to strengthen downtowns and solidify tax bases so the towns have adequate revenues to support community services such as schools, roads, and emergency services.

While a guiding principle for towns and counties should be to focus commercial development in existing centers, there are legitimate reasons to allow commercial development in undeveloped areas outside municipalities. Common-sense approaches should apply, and towns need to make sure that existing zoning does not impede compatible new operations and activities.

Emerging strategies that could help the traditional resource economy adapt to the changing global market and sustain itself over the long term include more sustainable agriculture practices; production and distribution of renewable energy, such as wind, solar, biomass, methane from livestock, and geothermal; and green jobs in former rural manufacturing plants converted to produce, distribute, install, and maintain green energy facilities and distribution networks. Most of these strategies will probably require changes to existing zoning and development codes.

Reuse of former industrial and commercial sites lets rural communities use their existing resources, preserve their heritage, and promote new economic activity. For example, this former mill in Front Royal, Virginia, is now a restaurant.

RESPONSE TO THE PROBLEM

Rural local governments are managing and encouraging commercial development in a variety of ways:

- Some local plans call for most commercial development to be located in incorporated municipalities, with a few exceptions.

- Some local governments sign formal intergovernmental agreements that implement these policies through zoning district regulations that do not allow commercial growth in outlying areas.

- Other jurisdictions that allow some commercial development outside towns have adopted design standards to help ensure that the new development respects rural character.

- Rural localities that have experienced commercial strip

development along entry corridors that lead into town centers from the surrounding areas are using corridor redevelopment strategies to convert aging shopping strips and underused parking lots into walkable, mixed-use destinations.[34]

Careful planning and close cooperation between towns and counties can help ensure that commercial development in rural areas strengthens the local economy while protecting the environment and the rural quality of life. This cooperation could include interjurisdictional agreements that articulate the value of emerging green industries. For example, entrepreneurs seeking to site wind farms and solar installations in rural areas are also considering rural locations for the related manufacturing and maintenance facilities, potentially providing new high-paying jobs.

Incentives can help direct commercial and industrial development to appropriate locations, like existing Main Streets or unused industrial, warehouse, or brownfield properties. Businesses might be more interested in reusing vacant properties when at least one property owner in the area has successfully converted a building back to productive use. Localities should make sure that in-town zoning allows, where feasible, the uses and services typically found in strip centers.

Many rural communities identify appropriate locations for expanded commercial or mixed-use development, including:

- Downtowns and adjacent commercial areas.

- Small commercial or mixed-use districts in residential neighborhoods near downtown.

- Commercial corridors, which have many buildings and aging sites that are underused or underperforming as retail or commercial businesses.

- Traditional industrial areas, agricultural service areas (often near railroads), and warehouse districts.

Downtowns and surrounding commercial districts usually have a variety of sites that can provide development opportunities. Commercial properties, including light-industrial or warehouse buildings, can be converted to mixed-use development with ground-floor retail or offices. Even small towns can have large industrial parcels ideal for transformation into commercial,

retail, or mixed-use districts. A financial feasibility analysis identifying appropriate potential uses can help the development community to understand the opportunities.

Small-town commercial corridors can suffer from aging, underused properties as well as competition from newer, outlying retail centers. They typically have greyfield (e.g., underused parking lots or shopping centers) and brownfield properties (e.g., former gas stations, dry cleaners, or industrial sites that might be contaminated), often at key intersections and within walking distance of residential neighborhoods.

Localities and business groups can map underused sites along major commercial corridors and evaluate their potential. Reusing these retail and service sites has several benefits:

- They are often large enough to be viable, mixed-use developments.

- Existing retail zoning might already allow commercial, residential, and mixed-use development.

- The connection to adjacent residential neighborhoods is often minimal, and new mixed-use development will be more compatible than existing commercial uses, helping to build neighborhood support for more compact development.

- Many older shopping centers were built at intersections, which can make redevelopment projects targets for enhanced or extended transit service or promising locations for future transit service, if none is currently in place.

Corridor redevelopment plans can be developed through a charrette, with government staff, residents, business owners, and elected officials creating a vision for the corridor and design concepts for specific sites. This approach can expedite redevelopment by providing general direction to potential developers, even before any longer-term transportation improvements are completed. These redevelopment plans could be used as guidance in a PUD process (see Chapter 3: Reform Rural Planned Unit Developments) or as design guidelines for a mixed-use project under retail zoning that allows residential uses. These corridors could also be receiving areas for TDR lands (see Chapter 9: Protect Agricultural and Sensitive Natural Areas). Local governments can assist in these types of projects by expediting design and review processes and by providing infrastructure financing for streetscape and utility upgrades.

---

34   ICF International and Freedman Tung Sasaki. *Restructuring the Commercial Strip: A Practical Guide for Planning the Revitalization of Deteriorating Strip Corridors.* EPA. 2010. http://www.epa.gov/smartgrowth/corridor_guide.htm.

Since residents of nearby neighborhoods sometimes object to redevelopment of corridors and downtown commercial districts, the town could adopt performance standards to measure and control noise, parking, lighting, and other neighborhood concerns. Similarly, the town could develop performance standards to encourage home businesses while minimizing any impacts. These standards should focus on the perceived impacts or concerns, like traffic or parking, rather than specific occupations or uses, to avoid the subtle bias that can sometimes arise. The community also needs a mechanism to determine when a home occupation or craft, such as tailor or woodworker, becomes a cottage industry. The same is true for farm-based businesses; a new "agricultural workplace" zone could allow commercial, light manufacturing, retail, or related uses on an owner-occupied farm, allowing home offices, on-farm sales, and agriculture-related industry (see Chapter 9: Protect Agricultural and Sensitive Natural Areas).

EXPECTED BENEFITS

• Directing commercial growth to towns and along corridors helps reduce scattered development in unincorporated rural areas.

• Active commercial centers and downtowns create a strong sense of community and bring shops, services, and employment.

• Development increases the tax base to support municipal services.

• Residents can walk or bike to stores and services, which could improve their health, save them money, and reduce greenhouse gas emissions and other air pollution.

• Redevelopment of aging corridors that do not fit with the town's desired character also helps avoid commercial development outside towns that detracts from rural character and scenic views.

• Capitalizing on public and private investment in renewable energy facilities in rural areas can generate jobs and tax revenues.

Photo courtesy of Jim Charlier

Encouraging commercial development, including small businesses, in the downtown strengthens the community and brings new activity to Main Street, as seen in Wells, Maine.

STEPS TO IMPLEMENTATION

1. Modest Adjustments

• Adopt a policy in county comprehensive plans to locate most commercial development in incorporated towns unless that development must be in an outlying location due to its use (e.g., processing agricultural products).

• Allow commercial development only in town influence areas or established unincorporated hamlets and crossroads villages with good connections to existing development, not in more remote locations.

• Direct state and local public works spending in ways that support and encourage activity in existing commercial areas in incorporated towns and discourage it elsewhere.

• If there must be commercial development in outlying areas, cluster it to create nodes instead of stringing it along the highway.

• Assess the support and customer base for additional retail development and match the zoning to the likely size of eventual build-out to help direct development toward preferred areas.

- Encourage new industrial activity in town influence areas by marketing sites adjoining rail stations and other locations where the community wants development. If the community is offering development incentives, it could give priority to projects that locate on these sites.

2. Major Modifications

- Prohibit rural commercial development in many county zone districts. Allow it only in service areas and locations designated in the comprehensive plan.

- Conduct a study of all available parking in downtown and commercial districts, and implement a parking management plan or "park once" district to encourage shared parking and to use parking more efficiently. When parking is developed at appropriate levels, uses can be more compact, and the community can add design amenities like streetscaping, which makes business locations more attractive.

- Conduct a planning study along an aging commercial corridor to identify key redevelopment sites and priority transportation improvements. Adopt any required zoning amendments or an overlay zoning code to allow and encourage redevelopment.

- Conduct a commercial market analysis for the downtown to identify commercial opportunities and needs.

3. Wholesale Changes

- Sign an intergovernmental agreement with towns in the region to share tax revenues from unincorporated commercial development.

- Assess road, safety, infrastructure, and other impact fees on rural commercial development to reflect the full cost of services and facilities needed for development.

- Assess the potential for renewable and alternative energy production and associated manufacturing and services. Determine appropriate locations, siting requirements, and regulations to encourage green industry and jobs.

- Identify any publicly owned land or buildings that are appropriate for commercial, industrial, or mixed-use development. Conduct a planning workshop to identify preferred uses and to spur redevelopment. Coordinate with local and regional business and industry organizations to develop a marketing strategy to recruit businesses.

- Consider creating a TIF district to encourage and fund downtown commercial development.

- Allow commercial development in outlying areas by special use permit only after requiring the developer to demonstrate the need for that service in that area. Adopt site and building design standards to ensure that any commercial development is in keeping with rural character.

PRACTICE POINTERS

- Joint town-county planning for commercial development in rural areas is usually essential to a successful implementation program.

- Encourage staff to investigate potential technical assistance and funding opportunities to reuse vacant properties and formerly contaminated sites.

- Many state departments of transportation and regional planning agencies have programs and grants to support revitalization of Main Streets and redevelopment of commercial corridors as long as vehicle movement and safety are also addressed.

EXAMPLES AND REFERENCES

Ballash, H. *Keeping the Rural Vision: Protecting Rural Character and Planning for Rural Development.* Washington State Community, Trade and Economic Development (Washington State Department of Commerce). June 1999. http://www.commerce.wa.gov/DesktopModules/CTEDPublications/CTEDPublicationsView.aspx?tabID=0&alias=CTED&lang=en&ItemID=974&MId=944&wversion=Staging

Challam County, Washington. "Lamird Report: Granny's Café." September 2006. http://www.clallam.net/RealEstate/assets/applets/PAPRlamird2-GrannysCafe.pdf.

St. Lucie County, Florida. "Towns, Villages, and Countryside" (Master Plan). 2008. http://www.spikowski.com/StLucieLDRrevisions-Ordinance06-017-AsAdopted.pdf.

www.ingramcontent.com/pod-product-compliance
Lightning Source LLC
Chambersburg PA
CBHW080648180526
45168CB00008B/3346